在 Arduino 帮助下掌握微控制器
第 2 版
（影印版）

Clemens Valens 著

南京　东南大学出版社

图书在版编目(CIP)数据

在 Arduino 帮助下掌握微控制器:第 2 版:英文/(荷)瓦伦斯(Valens,C.)著. —影印本. —南京:东南大学出版社,2015.9

书名原文:Mastering Microcontrollers:Helped By Arduino,Second Edition

ISBN 978-7-5641-5969-6

Ⅰ.①在… Ⅱ.①瓦… Ⅲ.①单片微型计算机-程序设计-英文 Ⅳ.①TP368.1

中国版本图书馆 CIP 数据核字(2015)189385 号

在 Arduino 帮助下掌握微控制器 第 2 版(影印版)

出版发行:东南大学出版社
地　　址:南京四牌楼 2 号　　邮编:210096
出 版 人:江建中
网　　址:http://www.seupress.com
电子邮件:press@seupress.com
印　　刷:常州市武进第三印刷有限公司
开　　本:787 毫米×980 毫米　16 开本
印　　张:24
字　　数:415 千字
版　　次:2015 年 9 月第 1 版
印　　次:2015 年 9 月第 1 次印刷
书　　号:ISBN 978-7-5641-5969-6
定　　价:74.00 元

本社图书若有印装质量问题,请直接与营销部联系。电话(传真):025-83791830

Table of Contents

1. Quick Start Guide

Because most people, including myself, lack the patience to read ten pages or so before connecting an Arduino board (or any other device for that matter) to a computer, here is how to see quickly if things work or not. What follows is a simple quick start guide. If, after reading this guide, you did not succeed in making the onboard LED flash, I advise you to read quietly at least the first few chapters to familiarize yourself with the Arduino platform and try again later.

1.1 Software Installation

Download the latest version of the Arduino software, you can find it on the *Download* page of the Arduino website (`www.arduino.cc`). Versions for Windows, Mac OS X and Linux are available, pick the one that suits you best. There is no point in downloading the source code or earlier versions, in this book I assume that you are using the Arduino software version 1.0.x where x is a number. Version 1.5.x can also be used, but while I was finishing this book (May 2013) it was still in the beta testing phase. We do not intend to modify the software, and so its source code does not interest us.

The Arduino software is provided in the form of a huge archive. To install it, unpack it somewhere on a hard drive (internal or external) or a USB key, using the standard method for your operating system. That's all; there is no nice graphical installer.

It may be necessary to install Java too if it is not available on your computer. Your operating system will let you know if you have to do this.

1.2 Installing the Hardware

The Arduino hardware installation is a bit more complicated than the software installation because of the diversity of the boards and the serial ports they offer.To facilitate reading this guide, I will limit myself here to Arduino boards equipped with a USB serial port. This family has two branches: boards with a USB serial port based on a chip manufactured by FTDI and boards with a USB serial port managed by a device of the brand Atmel. These last boards are the most recent, but all work very well.

Figure 1-1 - A selection of Arduino boards.

- Identify your board. This should not pose too many problems, since it is written on it in large characters.
- Connect the board to a free USB port of your computer. The computer may now emit a sound, display a message or do both (or do nothing at all) to inform you that it noticed that something was connected to it. An LED on the board should come on. If it doesn't, verify that the board is powered from the USB port. On some boards, like the Diecimila, you need to set a jumper to power the board from the USB port.

The next step depends on the operating system on your PC.

1.2.1 Windows (XP or Later)

Just as there are many roads that lead to Rome, there are several ways to install a driver in Windows. The techniques described below have been successfully tested by me on three different computers.

1.2.1.1 Board without FTDI chip (Uno, Mega 2560 or later)

Windows XP

After connecting the board, Windows will open the *Found New Hardware Wizard*. If you try to install the software automatically, Windows will not succeed, because you have to *Install from a list or specific location (Advanced)*. Click *Next*, check

Search for the best driver in these locations, check the box *Include this location in the search* and navigate to the subdirectory of the Arduino installation that holds the drivers. (The exact method to browse depends on the version of Windows on your PC.) Click *Next* again, Windows starts searching, finds a *Communications Port* and displays a message saying that the driver has not been signed. Ignore this warning by clicking *Continue*, then click *Finish*. If all goes well Windows displays a message informing you that the hardware has been installed correctly.

Windows Vista

A window entitled *New Hardware Wizard* will open. Click *Locate and install driver software (recommended)*. Windows now goes off looking for a driver, but, after a while, will return empty-handed and ask you for a disc with the driver on it. Since you do not have such a disc, click *I don't have the disc. Show me other options* and then click *Browse my computer for driver software (advanced)*. In the window that opens click *Browse* and navigate to the subdirectory of the Arduino installation that contains the board drivers. Click *Next* and then, in the warning window that appears, click *Install this driver software anyway*. Now wait for the message that indicates that Windows has completed the installation of a *Communications Port* and click *Close* to complete the process.

Windows 7

Connecting the board causes no visible or audible reaction of Windows. To install the board open the *Control Panel*, and then click *Device Manager*. Normally, in the list of devices, a small yellow warning triangle indicates that the Arduino driver has not been installed. Double-clicking the icon of the board will display its properties; click on the button *Update Driver*. In the next window, click on *Browse my computer for driver software* and navigate to the subdirectory of the Arduino installation that contains the board drivers. Click *Next*, and then, in the warning window that pops up, click *Install this driver software anyway* and wait for the message indicating that Windows has finished installing a *Communications Port*. Click *Close* and finish by clicking a second time on *Close*. Since the *Device Manager* is still open, use it to write down the number of the serial port assigned to the board.

1.2.1.2 Board with FTDI chip (Duemilanove, Nano, Diecimila, etc.)

After connecting the board to the computer, Windows either manages to install the driver all by itself (Windows Vista and 7, it may take several minutes for each driver and there are two; remember to write down the number of the USB serial port) or it proposes to have look on the internet. Refuse this proposal without cancelling the wizard and continue by selecting *Install from a list or specific location (Advanced)*. Select *Search for the best driver in these locations*, check *Include this location in the search*, navigate to the subdirectory *drivers\FTDI USB Drivers* of the Arduino installation and let Windows complete the installation. Two drivers are to be installed, so repeat these steps for each driver.

1.2.1.3 Finding the board's serial port number in Windows

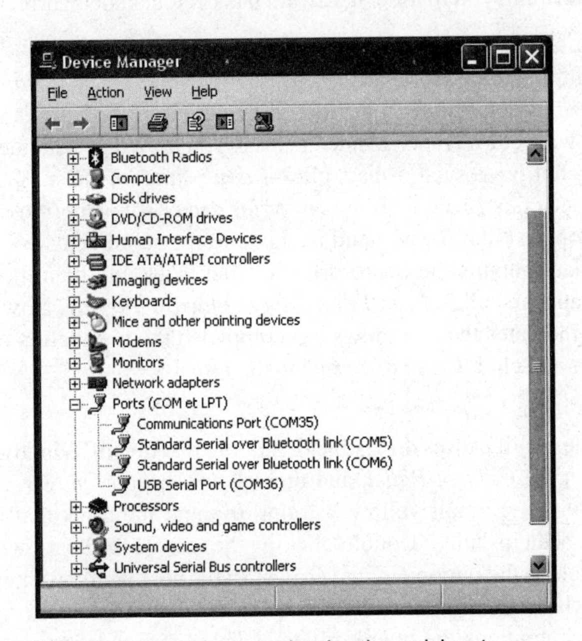

Figure 1-2 - The Windows Device Manager showing the serial ports.

Depending on the version of Windows and your configuration, there are at least two ways to find the number of the serial port that Windows has assigned to your board. Here is my preferred method:

Open the *Control Panel* and

Windows XP:
Launch *System*. A window entitled *System Properties* will open. Click the *Hardware* tab, and then click *Device Manager*.

Windows Vista:
Click *System and Maintenance*, and then click *Device Manager*.

Windows 7:
Click the *Device Manager* link.

Examine the *Ports (COM & LPT)*. If the board has been installed properly, you'll find a port with the same name as the one found by Windows during installation. In my case, an Arduino Uno was displayed as *Communications Port (COM35)* while the one with an FTDI chip on it was shown as a *USB Serial Port (COM36)*.

1.2.2 Mac OS X

1.2.2.1 Board without FTDI chip (Uno, Mega 2560 or later)

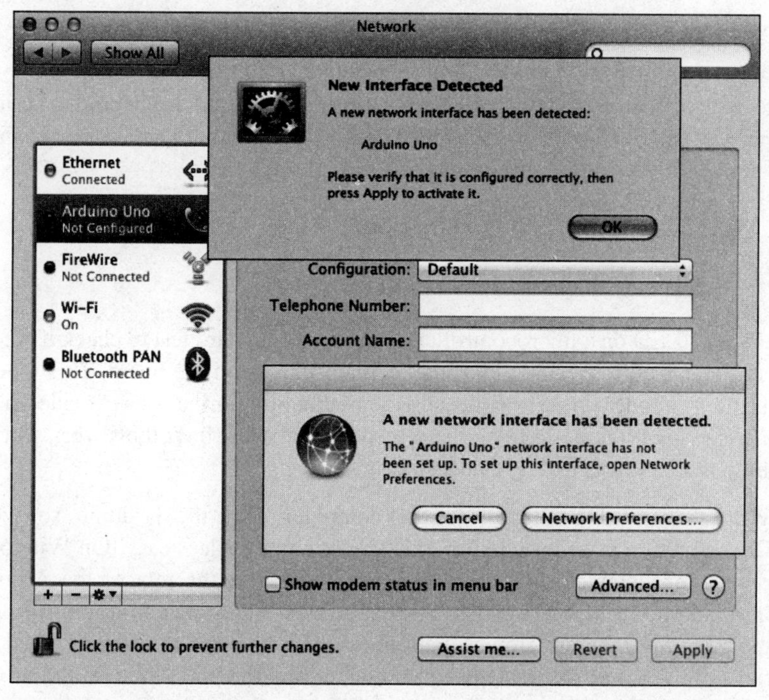

Figure 1-3 - On Mac OS X systems an Arduino board is considered a network interface.

Connect the board to the computer. Open *System Preferences*, then *Network*. A window may or may not open to announce that a new (network) interface named Arduino Uno has been detected. If it opened, click on *OK* or *Network Preferences...* to close it. The card is not connected or configured, but it will work. Click

Apply if possible or close the network preferences window in another way. This has been tested on Mac OS X version 10.5.8 ("Leopard", the pop-up window showed) and version 10.8.3. ("Mountain Lion", the pop-up window did not show).

1.2.2.2 Board with FTDI chip (Duemilanove, Nano, Diecimila, etc.)

According to the website arduino.cc, drivers for boards with a USB serial port based on an FTDI chip should be included in the Arduino package, but I have not found them (in *Finder* press the *Control* key and click on the package (or do a right mouse click), then select *Show Package Contents* to see what's in the big Arduino archive). If you need these drivers, consult the website of FTDI (`www.ftdi-chip.com`) to find out how to install them.

1.2.3 Linux

Because of the wide variety of Linux configurations, it is unfortunately not possible to explain in detail how to go about installing Arduino on this operating system. The easiest way is to visit arduino.cc for details. That said, if you are using Linux, chances are that you don't need my assistance to do this.

1.3 Hello World

Blinking an LED on a microcontroller board is the standard test to check if we are able to (technically speaking, not intellectually) program the board. This type of test is often called Hello World, because computer programmers have made a habit of printing this little message on the screen to verify that their tools work. On the Arduino platform this test is done as follows:

- With the board connected to the computer start the Arduino software. Depending on your installation, this can be more or less easy. On Windows, run the executable arduino.exe which is located in the root of the Arduino installation, on Mac OS X just click on the Arduino icon and on Linux you must run the program arduino (and not display it or run it in the terminal). If the Arduino software was up and running during the installation of the drivers for the board, it may not yet be aware of their existence and it will not list the serial port to which the board is connected. If this is the case, close and restart the Arduino software.
- In the menu, click on *Tools*, then click *Board* and select your board from the list that appears. If you are using version 1.5 or later of Arduino and if your

board is a model that can be equipped with different processors, such as the Mega, you must also select a processor from the *Processor* menu.

- In the menu, click on *Tools*, then *Serial Port* and choose the serial port to use. On Windows, this port is called COMx where *x* is the port number. Check the Windows Device Manager to find this number (see the paragraph above *Finding the board's serial port number in Windows*). On Mac OS X, this port is named /dev/tty.usbmodemXXX (board without FTDI chip) or /dev/tty.usbserialXXX (board with FTDI chip). On Linux it is /dev/ttyACMXXX or /dev/ttyUSBXXX where *XXX* is a number or a string that uniquely identifies the port.
- In the *File* menu click *Examples* followed by *1.Basics* to open the example *Blink*.
- Click on the button with the right-pointing arrow (*Upload*) to compile the example and program it into the board.
- The LED connected to digital output 13 should start flashing at a frequency of 0.5 Hz.

2. Introduction

2.1 Someone Knocks at the Door

Microcontrollers or MCUs are programmable components frequently used in an increasing number of applications. Thanks to their continuously decreasing costs and their ever increasing performance [Knock, knock, knock! Someone is knocking on the door? "Come in!"], today it is often more advantageous to use an MCU rather than a few discrete components.

"Oh hi, Dennis, Marilyn, what's up?"

Dennis and Marilyn are the editors who proofread and correct the manuscript of this book.

"Well Clemens, I find this introduction rather flabby.

- You do?

- Yes, it is so commonplace! For example, you could have challenged the reader and ask him how many microcontrollers he used since he got up this morning without even realizing it.

- Oh?

- In his clock radio, in the boiler of the bathroom, in the coffee machine, in the elevator, in the automatic door in his car, on the train, on the bus, etcetera. There must have been hundreds!

- He has an automatic door in his car?

- I forgot a comma.

- Okay. But... he takes the car, the train and the bus? He lives that far away from his work?

- I think you got the point!" [Dennis and Marilyn leave the room.]

Hmm ... Well, where was I? Ah yes, the microcontroller that is omnipresent. In short, even for flashing an LED a microcontroller is much better suited than conventional circuitry and the calculations needed to dimension passive and/or active components are nowadays often replaced by programming an MCU.

The development of microcontroller applications has long been the domain of specialist professionals, because the tools were expensive and difficult to use. All that has changed now and today creating a microcontroller-based design is within reach of the electronics enthusiast and beginner (even I can do it). Affordable microcontroller boards are all over the Internet and programming tools have become easier to use and are often offered for free. Indeed, the MCU has definitely become commonplace.

But beware. Commonplace does not mean child's play. While MCU manufacturers try to make this technology accessible to the masses, a good knowledge of the subject is still needed to transform a prototype assembled from development kits into a quality product.

2.2 Where Are We Going?

This book aims to introduce you to microcontrollers and microcontroller programming using the Arduino rapid prototyping platform. This platform, consisting of a cheap MCU development board and free programming software, allows me to introduce the various aspects of microcontroller programming in an accessible and inexpensive way.

I will present the most common microcontroller peripherals and illustrate their usage with small Arduino programs so you can understand how they work without getting lost in the details. Then we will look closer at each peripheral to learn how to use them outside Arduino, which enables you to continue your studies later on.

2.2 Where Are We Going?

This book is not in the style of "30 Funny Applications for Arduino". Even though there will be funny applications, it offers much more. At the end of the book, if you do not skip too much, you will be able to use any microcontroller and you can leave the Arduino cocoon to fly on your own wings.

To enjoy the chapters that follow, I suppose you have a minimum experience in computing and you know:

- How to use a web browser, your computer and its file manager;
- How to create folders and files;
- How to manipulate archives in various formats (especially those common for your operating system);
- How to find your downloads;
- What a USB port and a serial port are;
- What resistors, capacitors, transistors, etcetera are.

In order to concentrate on MCU programming, I will not explain all the details of the electronic circuits used in the examples. This book is not an introduction to electronics. Also I have drawn the circuit diagrams as circuit diagrams instead of colorful works of 3D art like you see (too) often in articles on Arduino. The illustrations in this book are all black & white anyway.

Happy reading!

3. Know Your Opponent

A microcontroller or Microcontroller Unit (MCU) is an integrated circuit comprising of a calculation unit, memory and peripherals. Unlike the microprocessor that is just a somewhat sophisticated arithmetic unit, the microcontroller does not require external components to run a program. MCUs that work stand-alone – not counting sensors and actuators like LEDs, displays, motors, etcetera – are becoming more numerous every day. Because their price continues to drop, microcontrollers find their way into all kinds of applications, from escalators and mobile phones (all types) to household appliances, toys, smart cards, etcetera. A modern car can contain up to fifty MCUs; several billion MCUs are sold every year...

3.1 A Short History of Microcontrollers

It seems that the first microcontroller or microcomputer was created in 1971 by engineers working at Texas Instruments (TI). I say "seems" because at that time several projects by various companies produced similar circuits. Still in 1971, the TMS1802NC is commercialized by TI. This was a calculator on a chip consisting of Read-Only Memory (ROM), a little bit of Random Access Memory (RAM), a calculation unit, a matrix keypad interface and a 7-segment LED display driver.

In 1976 Intel launched the 8048, an MCU that became a phenomenal success thanks to the fact that it was selected as the brains of the PC keyboard. The 8048 gave rise to a whole family, the MCS-48, replaced in 1980 by the 8051 and its family MCS-51. The 8051 core was licensed to a large number of chip manufacturers and it quickly became very popular. Even today many MCUs based on this core are available on the market (the well-known 8052 is an improved 8051).

Up to then MCUs generally existed in two variants: with factory-programmed ROM and with Erasable Programmable Read-Only Memory (EPROM) that the user, equipped with the right tools, could reprogram. A source of ultraviolet light is needed to erase the memory. In 1993 Microchip introduced the now legendary PIC16C84, the first MCU with Electrically Erasable Programmable Read-Only Memory (EEPROM), and Atmel launched the first 8051 with flash memory. These two memories are electrically erasable and it became possible to reprogram the MCU in-system, i.e. without removing the device from the circuit in which it lived and without special and expensive programmers.

Figure 3-1 - The power of lightning captured in a Leyden jar by Benjamin Franklin gave birth, 250 years later, to powerful microcontrollers in tiny BGA packages

Nowadays almost all microcontrollers have flash memory or another type of electrically erasable memory. MCUs with so-called masked ROM are being produced too because their programming is included in the manufacturing process, which make them cheaper in mass production.

Despite the arrival of 32-bit MCUs, boosted by the smartphone and tablet PC industry, 8-bit microcontrollers remain popular. 16-bits MCUs also exist, but they appear to be more discreet.

3.2 They are Cute, but What's Inside?

Regardless of the number of bits or the manufacturer of the microcontroller, almost all MCUs have a similar architecture. They combine on a single chip a processor, an oscillator, RAM and electrically erasable memory (EEPROM, flash, other) with many peripherals such as an interrupt handler, digital inputs and outputs, a multi-channel analog-to-digital converter (ADC), multifunctional counters/timers, communication controllers, a watchdog timer, etcetera. Depending on the size of the micro, these devices can be more or less powerful, and other features may be present such as a digital-to-analog converter (DAC), an analog comparator, a real time clock (RTC) or a direct memory access (DMA) controller. Specialized models may even contain radios or have, for instance, special functions for controlling motors.

3.2.1 The Processor

At the heart of the microcontroller sits the Central Processing Unit (CPU) or simply processor that executes the program and that processes the instructions. It contains the Arithmetic Logical Unit (ALU) which does the calculations; a controller or sequencer unit; some registers and an input-output unit that deals with the logistics of moving instructions and data to and from the ALU.

Most MCU processors are not especially good at math – microprocessors usually do much better – but on the other hand they are particularly handy with bits; they are real bit jugglers. This ability allows for efficient manipulation of I/O ports, but it is also useful for managing status, interrupt and device configuration bits. In fact, because the program memory is often quite small (always too small according to some), the MCU's processor is designed so that it can perform the most common tasks with a minimum of instructions.

The processor also contains a set of registers to facilitate calculations. Access to these registers is faster than accessing memory and so they are used to store intermediate results of calculations. They also help to keep on hand some important data or constants to speed up certain operations. Do not confuse the processor's registers with the peripheral registers that have completely different functions.

3.2.2 The Oscillator

A program is a sequence of instructions that are executed one after the other (we don't care about parallel execution here). This involves clocking and it is the processor's clock that is responsible for this task. The clock needs an oscillator, which often materializes as a quartz crystal and two small capacitors near the chip. When the accuracy of the clock is not too important, these three components can be replaced by a resonator or even a simple RC network. It is also possible to clock the MCU by an external oscillator, the one from another MCU for instance.

In reality, these external components represent only the tip of the iceberg, the oscillator consists of many more transistors than we imagine (try to create an oscillator with only a crystal and two capacitors, you'll discover soon enough that it is not easy). Moreover, many of these MCUs work perfectly fine without external components since they have everything they need on board. The on-chip oscillator's frequency is not necessarily very accurate, but it saves a few components and some space on the PCB, which is always interesting when you produce multiple (hundreds of) thousands of pieces of a system.

For high clock frequencies, manufacturers equip their devices with a Phase Locked Loop (PLL), making MCUs clocked internally at 80 MHz or more, commonplace.

The clock is also used by on-chip peripherals like timers and counters and also by communication devices that need it to send or receive bit streams at a well-defined rate.

3.2.3 Memory

The microcontroller likes to have more than one memory. Usually it has reprogrammable non-volatile (flash memory in most cases) to store the firmware (the program executed by the MCU), it has some random access memory (RAM) for storing data and variables and sometimes it also has a small non-volatile memory (EEPROM) for important data and configuration settings. The program memory is in general (much) larger than the RAM, which is OK, because most programs do not need that much RAM. As a consequence, memory-intensive applications such as smartphones or embedded Linux systems cannot use just any MCU and specialized models capable of working with large external memories are available for this kind of application.

Even if the processor has a 32-bit core, memory is often accessible byte by byte (one byte equals eight bits). Depending on the width of the ALU, memory can also be read or written in words of 16 or 32 bits, or words of another size.

Each memory position has a unique address, like the apartments in a building or the rooms in a hotel, and to access a data word – that may be a byte – it is necessary to specify its address. The number of bits of the address determines the number of data words the CPU can access. A 16-bit address can address 2^{16} = 65,536 words or 64 kilowords, a 32-bit address can address 2^{32} = 4,294,967,296 words, or 4 gigawords. Traditionally, in computer speak, kilo equals 1024 and giga equals 1,073,741,824. However, since 1998 this is no longer valid and 1024 now equals one kibi. A memory of 65,536 bytes therefore has a size of 64 kibibyte (or kibi) and a 1 TB hard drive can contain 10^{12} bytes and not 2^{40} bytes!

To improve performance some MCUs are based on a structure known as modified Harvard architecture. In such an MCU the processor can simultaneously access program and data memory. These memories may use different word sizes.

3.2.4 Interrupts

Imagine: you're set in your living room quietly trying to program a microcontroller when suddenly your phone signals an incoming text message. What do you do? You interrupt your exciting activity to read the message. Why? Because you are afraid to miss something important. By behaving this way, you responded to an interrupt request that you granted a higher priority than the task you were executing. For an MCU it is the same. Most of the time it quietly executes the main program, but every once in a while it is interrupted by a peripheral that needs a little attention. Contrary to your text messages, MCU interrupt requests are always important, because we do not bother an MCU with futilities. So, when it receives an interrupt, the MCU drops everything to respond as quickly as possible. It may be an alarm that triggered the interrupt request or some important data that came in and if the MCU responds too slowly, the data will be lost or the oil refinery may explode.

Where the first microprocessors and microcontrollers offered at best two or three interrupt sources, modern processors have lots of them. A specialised interrupt controller which occupies many pages of the MCU's user manual manages these sources. Since an MCU has many peripherals that all may generate one or more interrupts, you can easily understand why an interrupt controller can come in handy. It allows you to enable and disable interrupt sources or impose priorities. Some MCUs limit the number of interrupts a program can use, therefore selecting the peripherals that are allowed to generate interrupts and assigning their priorities require serious reflection from the system designer (you).

Usually the reset signal is also considered an interrupt. Several reset sources are distinguished like the external reset signal (a pushbutton for instance), a reset due to a (temporary) drop of the supply voltage (brown out) or a program runtime error.

3.2.5 Input/Output Ports

These are the ports of the MCU peripherals that are connected to the pins of the chip, and that provide the interface to the real world. A microprocessor provides access to a data bus, an address bus and a control bus whereas a microcontroller only has multi-purpose input/output (I/O) pins (in the literature one speaks of General Purpose Input Output or GPIO) organized in ports of often 8, 16 or 32 bits. The ideal microcontroller only has ports (and consumes exactly 0 W), but it would need a lot of them to provide access to all of the peripherals. MCU manufacturers therefore prefer to combine (multiplex) several functions per port to limit the number of pins of the chip, with the consequence that it is impossible to use all peripherals simultaneously. It is the job of the system designer (you) to decide how to use the MCU ports in the best way.

3.2.6 Analog-to-Digital Converter

To interact with the outside world where signals are usually analog, the microcontroller uses an Analog-to-Digital Converter (ADC). Most MCUs equipped with such a peripheral only have one of them, even though the datasheet seems to say otherwise. In reality, the ADC is wired to a multiplexer that connects only one out of several inputs (channels) to the ADC. Converting all ADC channels at the same time is therefore not possible. A workaround is sometimes provided by allowing chaining of the channels so that they can be processed automatically one after the other in a loop.

ADCs can be designed in several ways, but those found in MCUs are often of the successive approximation (SAR) type. This type of ADC refines the conversion result in several steps prior to releasing it to the user. This technique is not the fastest, but it has the advantage that it doesn't require too much chip real estate. Also, slow is not that slow as conversion times are measured in microseconds; usually they are more than fast enough for most of your applications.

At the time of writing, 10-bit ADCs are the most common, but 12-bit ADCs are on their way.

3.2.7 Digital-to-Analog Converter

MCUs that have a Digital-to-Analog Converter (DAC) are not very common. The reason is that DACs are transistor hungry and demand a lot of space on the chip, a little too much to the liking of chip manufacturers who prefer to use this space for other functions. To address the absence of a DAC, manufacturers have found a compromise: a peripheral capable of producing a digital signal easy to convert to an analog signal. This compromise is called Pulse Width Modulation (PWM) and it consists of producing pulses of variable duration or width. The signal's period, i.e. the time between pulses, is fixed, but the duration of each pulse is proportional to the desired value of the analog signal. The pulse width expressed as a percentage of the period, the duty cycle, ranges from zero (0%, no pulse at all) to one hundred (100%, the pulse occupies the full period). A resistor and a capacitor wired as an integrator (or a more sophisticated filter) can calculate the average value of the pulse durations that corresponds to the desired analog value.

However, in many real-life applications it is not necessary to calculate the average of the PWM signal, because the inertia of the controlled device (or observer) does it naturally. For example, an LED controlled by a PWM signal will flash, but if the pulse rate exceeds a certain frequency, your eyes will not be able to distinguish the individual flashes and you will see light of constant intensity. By varying the pulse width of the PWM signal, you can vary the light intensity.

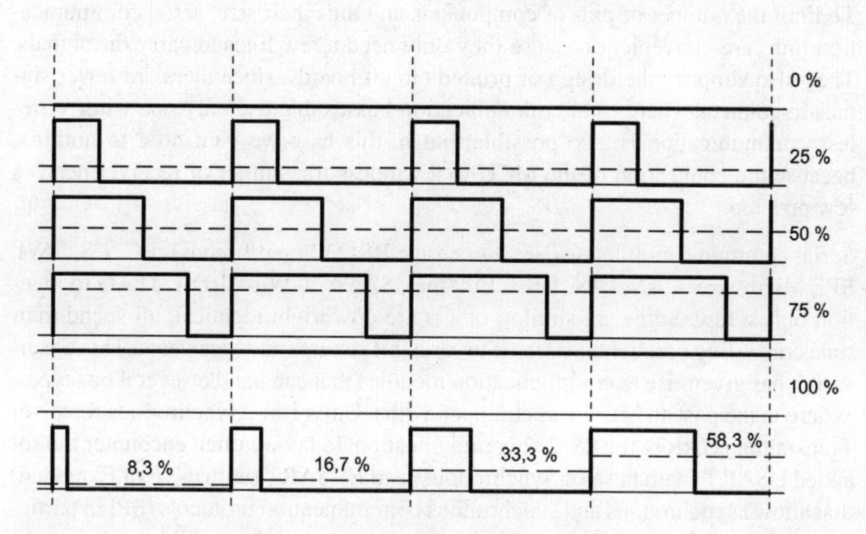

Figure 3-2 - Averaging a pulse width modulated signal with an RC network or another low-pass filter allows the creation of analog signals

Controlling something with a PWM signal is a very common technique. Not only can you use it to adjust the brightness of an LED, it is also possible to modulate the contrast of a liquid crystal display (LCD) or the speed of a motor. In the case where a voltage is really necessary, a PWM signal can be filtered to obtain it. Connecting an external DAC chip to the MCU is, of course, also an option.

3.2.8 Communication Modules

In many systems, the microcontroller must exchange data with one or more other devices. Data is composed of bits, the smallest amount of data occupies one bit and to transport it you need a channel. A byte contains eight bits and eight channels are needed to transport these bits at the same time. When several bits are exchanged at the same time over multiple channels, we speak of a parallel communication link. You can also send the eight bits of a byte one by one over a single channel and in this case we speak of a serial communication link. Serial communication needs fewer channels than parallel communication, but it is slower. If the time required to send a byte or a bit is the same, sending the eight bits of the byte sequentially takes eight times longer than sending them in parallel. On the other hand, a serial connection uses fewer channels than a parallel connection, eight times less in our example.

To limit the number of pins of components and thus their size, serial communication links are convenient because they only need a few lines to carry the signals. They also simplify the design of printed circuit boards, since there are fewer signals to connect. There exist communication busses that use only one wire. Wireless communication is also possible, but in this case we gain little to nothing, because the connection of the MCU to a wireless transmitter or receiver needs a few pins too.

Serial communication busses are numerous: RS-232 (and friends), I²C, I²S, TWI, SPI, Microwire, CAN, LIN, USB, Ethernet, SATA, 1-Wire, IrDA, JTAG to mention only a few. Some are similar, others are downright identical, all spend their time converting parallel bit streams to serial bit streams and vice versa. This observation has given rise to communication modules that can handle several bus types. Where in the past an MCU was equipped with a Universal Asynchronous Receiver Transmitter (UART) for RS-232 communication, today we often encounter the so-called U***S***ART (with the *S* of Synchronous) and ***E***U*S*ART (with the *E* of Extended) that allow asynchronous and synchronous communication protocols (SPI in particular, sometimes also I²C). Other buses are inherently incompatible and are provided as separate modules (CAN, USB, Ethernet, etcetera). Therefore, beware when a datasheet of a microcontroller mentions, for example, six serial ports, four SPI ports plus four I²C ports. When you read the small print you will realize that in reality only two of these communication busses may be used simultaneously.

3.2.9 Time Management

Measuring time is one of the favorite occupations of microcontrollers in the majority of applications and therefore they need timers. To meet this need, MCU manufacturers equip their products with several timer and counter peripherals. Because these modules are often very flexible, they can be difficult to understand. Their flexibility allows them to measure time lapses and pulses lengths, produce pulses with precise durations and, for instance, generate PWM signals. For good accuracy the timers/counters often have a width of 16 bits; sometimes they can be split into two 8-bit counters or, alternatively, be coupled to create a 32-bit counter. The timer clock signal can come from many sources, with or without prescaler, and they cover timing ranges from microseconds up to seconds, minutes or more.

With the increasing popularity of operating systems for microcontrollers, chip manufacturers have started to offer MCUs with a special timer, the system tick timer, to provide the 1 or 10 ms heartbeat that is so important for the operating system to execute its periodic tasks.

For applications that need to know the time and date and keep this information even when switched off or in standby mode, the MCU can be equipped with a Real-Time Clock (RTC). Such modules require a watch crystal[1] and sometimes their clock frequency can be fine-tuned in software to improve the RTC's accuracy.

Finally, to bring it back on track in case it got stuck in an endless loop or if for some unknown reason it lost control while executing a particularly difficult jump, the MCU is equipped with a so-called Watchdog Timer (WDT). This timer must be cleared every *x* milliseconds by the microcontroller. If the latter, lost, stuck or asleep, fails to do this in time, the watchdog will bark and the MCU will be reinitialized.

3.2.10 Other Peripherals

Depending on the applications targeted by the microcontroller manufacturer all kinds of peripherals can be integrated, even analog functions. There exist MCUs with Direct Memory Access (DMA) modules, LCD controllers, keypad or touchpad drivers, radios, analog comparators, parallel ports, etcetera. Also, to facilitate the development of a program or application, MCUs are often equipped with one or more peripherals that allow debugging of the program. With such a module it is possible to inspect the registers and memory, modify values of variables, and execute instructions manually to observe their effects and so on. Such a module is a powerful tool for finding and fixing your programming mistakes. Debugging peripherals generally communicate through a port compatible with the standard defined by the Joint Test Action Group (JTAG) or a variant. In general this port can also be used to program the flash memory of the MCU. A special – and often expensive – adapter is necessary to take full advantage of these debug and programming features.

3.3 Tools

Developing a microcontroller-based application requires a certain number of tools. I highly recommend the oscilloscope, the multimeter and a good (laboratory) power supply for the development of the electronic circuit surrounding the MCU and for debugging the program the MCU has to execute. To write this program, you must choose a programming language. The words you write in this language

1. A quartz crystal tuned to 32,768 Hz. This frequency can be used to obtain 1 Hz by dividing it 15 times by 2 (2^{15} = 32,768), which is easy to do with 15 flip-flops.

cannot be used directly by the MCU; tools are needed to transform it into code the MCU's CPU can execute. To load the executable code (or just executable for short) into the MCU, you will need a programmer. A desktop or portable computer to do all this is, of course, essential.

Let's start by investigating the tools that help you create the program: the programming tools.

3.3.1 Programming

The program executed by the MCU is a list of codes that describe every single step, even the smallest, necessary to achieve the result. This code is gibberish to a human being. Because it is quite difficult to write a working program in a code that is hard to understand, MCU manufacturers propose a low-level programming language called assembler. This language is fairly close to the executable code but a bit easier to read. I talk about assembler in the singular, but every MCU family has its own dialect. A software tool, also called assembler, transforms the program written in assembler into executable code.

In assembler, the programmer must specify everything. For example, to add two variables you must tell the MCU to fetch the two variables from memory one at a time, tell it to add them and tell it to store the result somewhere. This is tedious work and the risk of errors is large. In addition, a program written in assembler is difficult to understand and modifying or extending it when you did not write it or when you haven't touched it for some time is far from easy. Another major disadvantage of assembler code is its lack of universality or 'portability' as the computer programmer calls it. A program in assembler only works on the MCU for which it was written. To be executed by an MCU from another manufacturer or even by one from another MCU family from the same manufacturer, everything must be rewritten (an alternative might be to use an emulator).

Over the years, the numerous inconveniences of executable code and programs in assembler added to the fact that programs are getting bigger all the time has prompted computer scientists to design more structured, more readable and more powerful programming languages. This has led to the emergence of so-called "high-level" languages as Fortran, C, Pascal and BASIC. The goal of all these languages is to facilitate the creation of a working program and to improve its quality by providing libraries of tested functions and a well-defined structured syntax. A special programming tool, the compiler, transforms the high-level language in assembly language which in turn is transformed into executable code. Such

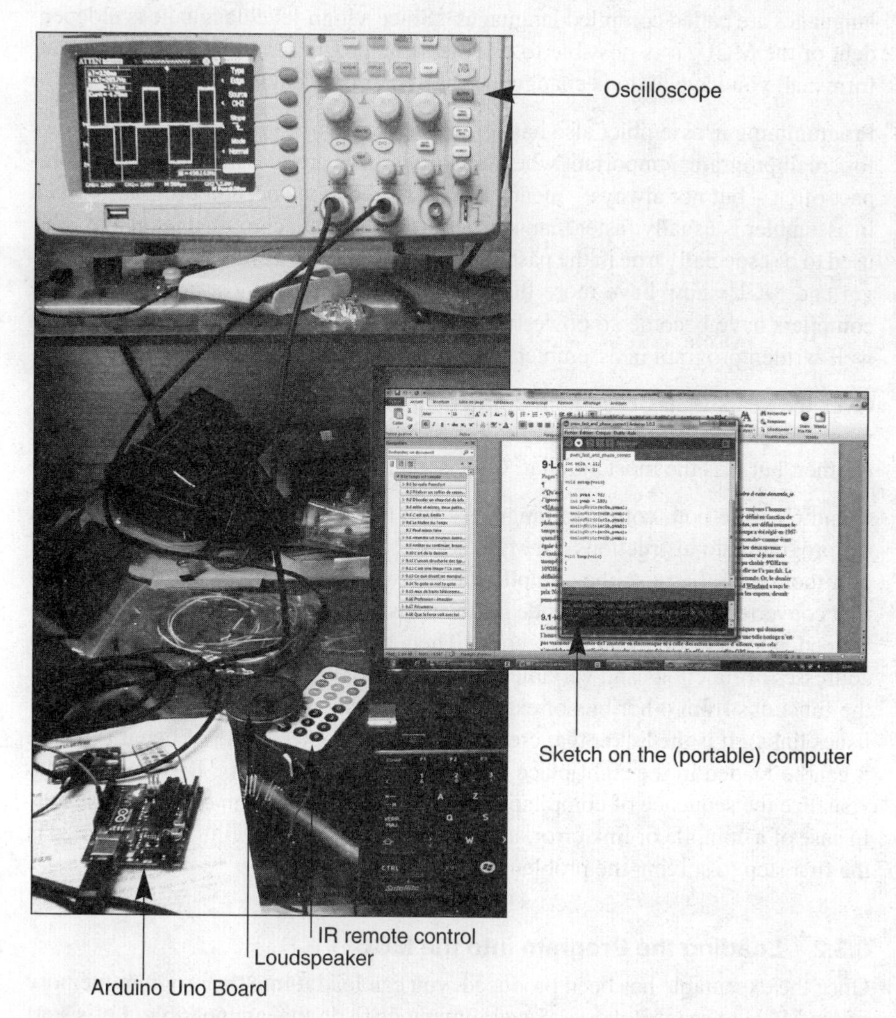

Figure 3-3 - From now on, avoid cleaning up your work environment. According to German scientists, disorder stimulates creativity.
To celebrate the start of your new vocation, buy yourself an oscilloscope (new or used). Believe me, it will change your life.

languages are called compiled languages[1]. Since a high-level language is independent of the MCU, it is possible to use the same program on different MCU platforms; all you have to do is change the assembler (the language and the tool).

Programming in assembler also has some advantages like its efficiency that allows for small programs, important when memories sizes are measured in bytes. Compact often – but not always – means a higher execution speed, a program written in assembler is usually faster than a program written in a compiled language. This used to be especially true in the past, but, as we all know, *the times they are a-changin* and MCUs now have more than enough memory for most applications and compilers have become so powerful that they can produce code comparable to a well-written program in assembler.

Over the years C has become the programming language of choice. Of course, many developers prefer to program in assembler, BASIC, Pascal, Fortran or yet another, but C is the most popular. C++ is gaining in importance.

C and C++ are both compiled languages and a compiler is required to transform the program into instructions the MCU's CPU can execute. In fact, we must speak of a tool-chain, because the compilation process is done in several steps. C++ is first converted to C code; this code is then translated into assembler which is transformed into bits of executable code. Then a linker ensures that the memory addresses of functions and variables in these bits of executable code are known by the functions from other bits of executable code that need them (the linker establishes links). It is the linker that creates the so-called executable. When it is ready, it can be loaded in the right place in the memory of the MCU. It is important to visualize the sequence of compilation steps, because errors can occur at all levels. In case of a compile or link error, understanding at which level it was generated is the first step to solving the problem.

3.3.2 Loading the Program into the MCU

Once the executable has been produced, you can load it into the program memory of the MCU to see if it works. Several methods to do this are possible. Let's start with the all singing, all dancing programmer. This type of tool is expensive, but can be used to program different MCUs from various manufacturers. These tools are fast and offer a lot more functions than a simpler model, such as a unlocking a misconfigured MCU. In addition to their price, they have another drawback: you have to extract the MCU from its circuit board, put it in the programmer, program

1. BASIC is a particular language because it can be compiled or interpreted (meaning that it is translated on the fly during program execution).

it and then put it back in the circuit. This quickly becomes tedious and such tools are reserved for production where a large number of components is programmed before being inserted into the circuit board.

Since the introduction of flash memories and the like, the components can be programmed without being removed from their board. This technique is called In-System Programming (ISP), also known as In-Circuit Serial Programming (ICSP) or Device Firmware Update (DFU). In-system programming can be achieved in several ways:

- Over a JTAG or similar interface. In this case, the programmer can access (almost) directly the MCU's memory and registers, the CPU is disabled, and it is the programmer that is in charge;
- Over a special ISP or ICSP port. The programmer communicates with a special on-chip peripheral that can directly access the (program) memory; the processor is disabled. The programmer sends the data and the addresses where the data should go, the ISP peripheral takes care of storing it in program memory;
- The programmer communicates with a software module that uses the CPU to access memory. The programmer sends the data with their addresses and the processor puts everything into memory. This technique works only when the processor has access to special instructions to write its program memory.

The last option opens the way for devices that the user can update by him- or herself without the need for special tools. Two methods are possible: either the device features a firmware update function that the user must manually activate or the device passes at start-up through a special so-called boot phase during which a small program is run that loads the application. The boot loader scans a communication port to see if a firmware update is desired. If this is the case the boot loader burns the new executable in memory and then runs it.

3.3.3 Debugging

Very rarely does a program work perfectly fine the first time it is run. It is much more common to go through a few debugging cycles before the final result is satisfactory. Your debugging experience depends on the depth of your pockets. With the right tool, the In-Circuit Debugger (ICD), it is possible during program execution to stop and restart the program whenever you want, at the times and places you want, or to walk through the code and inspect variables and registers at leisure.

Figure 3-4 - Well equipped or not, if you're a poor programmer, things will not work.

Even better still is the In-Circuit Emulator (ICE) that doesn't even need the MCU. It replaces the MCU and is controlled by the computer. Fast and transparent, it is the ultimate in debugging for the professional developer.

For the less wealthy there is the simulator. This software tool allows you to see what the MCU should be doing in theory, because it simulates and does not communicate with the real hardware. The simulator can be useful to develop an algorithm without having to reprogram the MCU fifty times. Relatively simple but slow in-circuit debuggers (debug pods) that offer one or two breakpoints are an interesting alternative for the developer who has to justify his/her investments to his/her spouse.

The good news is that in most cases bugs can be found with simple means: an LED, a serial port and your brain. If you have one – an oscilloscope, I mean – do not hesitate to enlist it in the debugging process as it can be very useful. Use (one of) the serial port(s) of the MCU to send informative messages about the status of the program to the PC, one or more LEDs are useful for visualising the levels of selected pins, and the oscilloscope can display fast signals. Free ports can be used to output status signals. Look closely at your program, put yourself in the place of the MCU, what would you do in a given situation? Think carefully before investing.

4. Rapid Prototyping Italian Style

Arduino, from Italy, is a rapid prototyping platform for microcontroller applications. This means that it is a set of tools developed to facilitate the design of microcontroller-based circuits without wasting too much time on learning the ins and outs of the platform. With Arduino, you too can do "microcontrollers" and "embedded electronics". *Now I'm gonna make you an offer you can't refuse*: get yourself one of those cheap Arduino boards, install the free software tools, keep your evenings free and I will teach you how to use it all.

4.1 The Godfather 1, 2 and 3

Arduino is related to the open-source techno-popularizing projects Processing from which it borrowed the programming environment and Wiring that provided the foundations of the code libraries.

Processing "*is an open source programming language and environment for people who want to create images, animations, and interactions. Initially developed to serve as a software sketchbook and to teach fundamentals of computer programming within a visual context, Processing also has evolved into a tool for generating finished professional work.*" (source: `www.processing.org`)

Wiring, Italian like Arduino, is a programming framework for microcontrollers "*created with designers and artists in mind to encourage a community where beginners through experts from around the world share ideas, knowledge and their collective experience.*" (source: `www.wiring.org.co`)

Arduino (`www.arduino.cc`) was conceived at the same school as Wiring and is a simplification of the latter, which in itself was already quite easy to use. Originally Wiring was not open-source, but under the pressure from the Arduino team, Wiring gave in and published its source code. Then the Arduino team could get going.

The goal of Arduino is to make microcontrollers accessible to students, non-specialists, artists, designers, enthusiasts and all those who are interested in the creation of interactive objects and environments, but who do not have (well-developed) electronics and/or programming skills and who do not always have a lot of money. Therefore the Arduino team decided that their board should not cost more than \$30 (€25), which was indeed the price of the first board. Since then prices have gone up marginally to reach about \$35 (€30) in 2013.

The first Arduino boards were more basic than the Wiring boards, but over time the Arduino hardware became more elaborated, whereas Wiring extended its range by simplifying their boards. Now (in 2013), Wiring S and Arduino Uno boards are very similar.

Despite the equivalence of the two projects, Wiring did not meet with the same success as Arduino. From the early beginning the Arduino team has done everything to spread their word by publishing all source code, circuit diagrams, printed circuit board designs and detailed documentation for free, whereas the Wiring team tried to control everything. In the Fall of 2012 over 600,000 Arduino boards had been sold worldwide.

Arduino libraries are based on those from Wiring and the Integrated Development Environment (IDE), based on the one from Processing, is almost identical to the Wiring IDE. The only striking difference is its color: orange for Wiring and light blue for Arduino. Processing uses shades of grey for its IDE with looks very familiar to the two others. Other projects have started to use the Processing IDE as well and we can almost speak of an emerging standard.

The Wiring IDE supports Arduino hardware, but the Arduino IDE does not know about Wiring hardware. Initially Arduino limited itself to a few microcontrollers from Atmel's 8-bit AVR family while Wiring openly flirted with microcontrollers from for instance Microchip and Texas Instruments and also with MCUs based on an ARM core. Arduino too has been transformed into a multi-MCU platform, but a little "unbeknownst to its own free will". The Arduino philosophy has been embraced by many enthusiasts who have ported it to different MCU platforms from various manufacturers. It took a few years to modify the Arduino IDE to allow it to include compilers for other microcontrollers as well, but with version 1.5 and the introduction of the Arduino Due that sports an ARM Cortex-M3 based MCU, this goal has been reached.

4.2 Pasta, Cheese and Tomato Sauce

The Arduino platform is built on three pillars:

1. The hardware consisting of a collection of microcontroller and expansion boards. Circuit diagrams and printed circuit board designs are available for free;

2. Software comprising of programming tools and an extensive library of high-level functions. Everything is free, open-source and multi-platform;

3. Distribution and communication in the shape of a website through which the Arduino hardware and software are made available to interested parties. The address is `www.arduino.cc` (`arduino.cc` also works). For the curious, `.cc` is the internet country code top-level domain reserved for the National Territory of Cocos (formerly Keeling) Islands, an Australian territory. This top-level domain was probably chosen for its low cost, but I digress.

The website is *the* official reference for everything that concerns Arduino. The website contains software and hardware updates and all the information necessary to use it all. If you are looking for a reseller of authentic Arduino hardware, it is here that you can find one. A forum to ask questions that are not answered in this book is also available. The site is in English, but there are sub-forums accommodating other languages too.

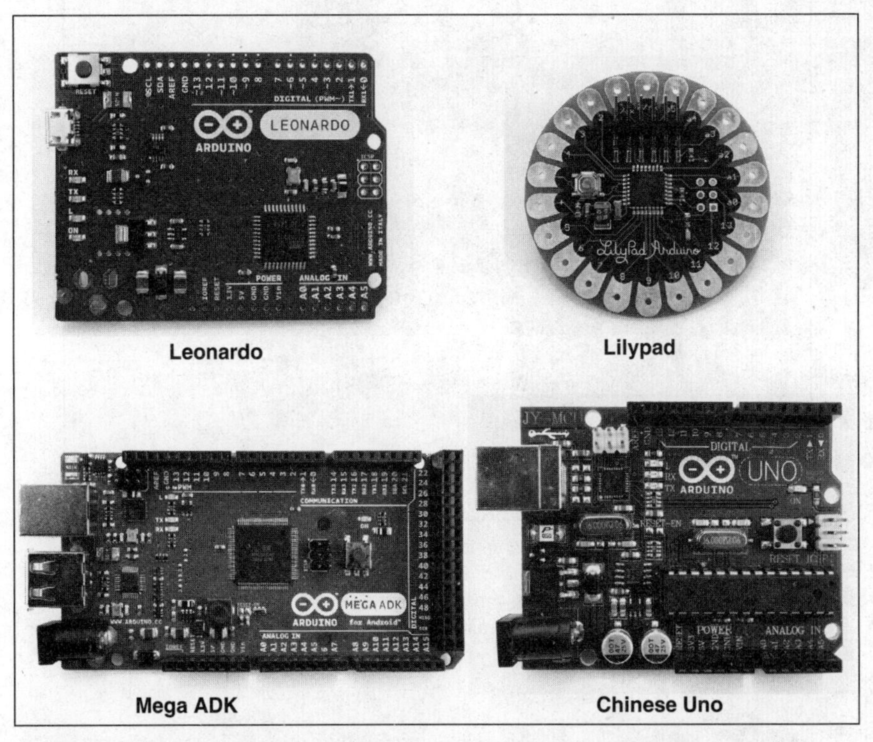

Figure 4-1 - Some less common Arduino boards: the Leonardo (without connectors) is equipped with an ATmega32U4 and targets USB devices; the daisy-shaped Lilypad aimed at wearable applications; the Mega ADK for communication with Android devices and a Chinese Uno clone recognizable from the missing "Made in Italy" statement.

On the hardware side of things, Arduino is a microcontroller board with a USB port through which it can be programmed. Several models exist, but the most common are the small Uno-shaped boards (Diecimila, Duemilanove, etc.) and the larger Mega 2560-style boards (Mega, Mega 1280, Mega ADK, etc.). There are others like the Mini, the Micro, the Nano or the round, daisy-shaped Lilypad intended for wearable applications, supposed to be used in clothing. The Arduino Due is different from all others because it is equipped with a 32-bit microcontroller instead of an 8-bit one and it is therefore much more powerful than the others[1]. The Yún, a combination of an Arduino Leonardo extended with a Wi-Fi system-on-a-chip running Linino (a MIPS GNU/Linux based on OpenWRT) has just been released.

Figure 4-2 - A collection of Arduino compatible add-on boards, or Shields. Shown are a Wi-Fi/Bluetooth shield, an Ethernet shield (white), a graphical display shield with a mobile phone display on it, a shield with a 4 x 4 matrix keypad, a monome *10h* with 16 large white LEDs (see `www.monome.org`) and left from the center an experimental shield that we will build in Chapter 8.

1. The Arduino Due is not treated in this book. Not only did it arrive when I had almost finished this book, but it also makes things more complicated. A bit of Arduino's simplicity has been traded in for a more powerful microcontroller.

The Arduino board can be programmed in C, C++ or assembler using open-source tools available for Windows, Mac OS X and Linux. The hardware is also open and anyone can make their own Arduino board. Schematics and printed circuit board (PCB) designs can be downloaded for free and instructions on how to use it all are published on the website. If you buy a cheap Arduino Uno from an unofficial dealer, it is possible or maybe even likely, that you will receive a clone produced somewhere in China. A genuine Arduino board has "Made in Italy" printed on it, clones do not. Even though all the information to make an Arduino board is available for free on the internet, you are not authorized to make thousands of them. Marketing a compatible board is of course allowed as long as it is not identical.

The Arduino board is equipped with expansion connectors that can accommodate compatible add-on boards. Such an extension board is called a Shield and there are hundreds of them out there, designed by Arduino users all over the world for all kinds of applications. Several official Arduino-branded shields are also available, such as the Ethernet Shield, the Wireless SD Shield, the Motor Shield or the perforated Proto Shield for prototyping.

4.3 Base Ingredients

An Arduino board is fairly simple – until the appearance of the Due – since it is basically an 8-bit microcontroller with expansion slots to which the in- and outputs of the microcontroller are connected. Such a board is also known as a break-out board and they exist for many components, especially those that are too small to use on a prototyping board or a breadboard. ATmega*X* MCUs (where *X* is a number) belong to the megaAVR family from Atmel. The first Arduino boards were equipped with an ATmega8, the pre-Uno boards sported an ATmega168 whereas the Uno has an ATmega328. The Mega 2560 is based on an ATmega2560, and the Mega 1280 on an ATmega1280. All these MCUs are more or less compatible; the only things that differ are the pin count, the memory sizes and the on-chip peripherals. The Arduino Due is different from all other boards because it is equipped with an Atmel SAM3X8E 32-bit microcontroller based on an ARM Cortex-M3 core.

The microcontroller on an Arduino board contains a small boot loader, which allows loading a new application into the MCU through a serial port without special tools and without overwriting the boot loader. With this little piece of software, reprogramming the board is easy. Unfortunately, modern computers no longer have serial ports, only USB ports (besides of course other unusable ports

Figure 4-3 - Drawing of the Arduino Uno board showing the most important components. The board was designed for educational use, which is why it has this atypical outline that allows to explain to students how to position the board without requiring any knowledge about what exactly is on the board.

like HDMI, VGA, etc.), and an adapter that converts a USB port into a serial port is required. To avoid having to buy a converter on top of the board, the Arduino designers decided to include one on the board.

During development the USB port of the PC is also used to power the board, but when the application is finished and the board has to survive on its own, it must be powered in another way. An on-board voltage regulator offers some freedom in the choice of the external power source.

Besides a pushbutton to restart the MCU (reset) and some configuration jumpers, the board does not contain much else. An LED is often present – it is used in several Arduino programming examples – and it can be helpful to determine if the Arduino environment (board + programming software) is functioning as it should.

The MCU's clock is based on a 16 MHz crystal oscillator. This is an important detail, because the boot loader relies on this value to adjust the baud rate of the serial port. If you overclock the board (the MCU is specified up to 20 MHz) the boot loader will have communication problems.

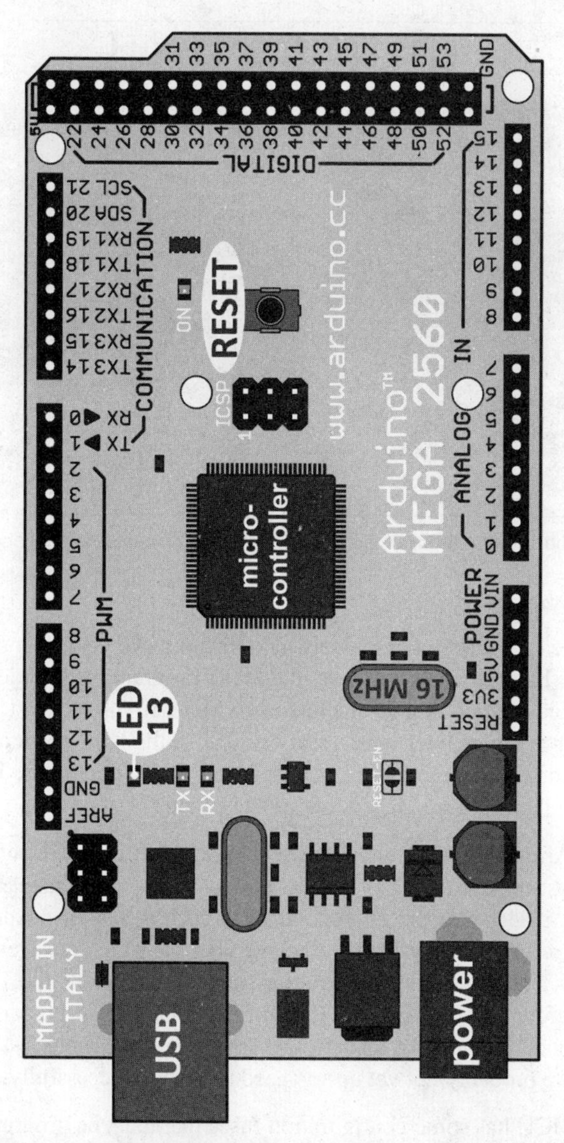

Figure 4-4 - The Arduino Mega (2560) is longer than the Uno because of the extra connectors needed to access the additional I/O that the ATmega2560 offers compared to the ATmega328 of the Uno. The Mega is compatible with the expansion boards for the Uno.

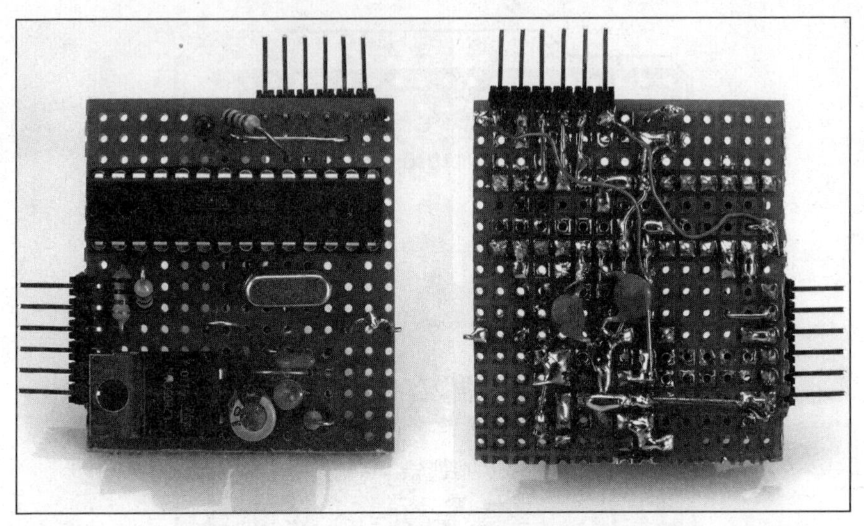

Figure 4-5 - A homemade Arduino compatible board, component side (left) and solder side (right).

That's all there is to know in case you would like to make your own Arduino compatible board. Basically, all you have to do is to program the boot loader into the ATmega device of your choice, clocked at 16 MHz. If you have a USB-to-serial-port converter or a computer with a real serial port, the USB portion of the board can be omitted. An RS-232 to TTL level adapter is essential in the last case (and maybe also in the first case, depending on the adapter).

Building an Arduino compatible board seems easy, but there are some complications. The most important one is the fact that an AVR programmer is required to load the boot loader into the microcontroller. Such a tool is inexpensive, but it means an extra piece of hardware. Another obstacle may very well be the boot loader that is not necessarily available as an executable for the AVR of your choice. Therefore use the ATmega328 (or ATmega168) or prepare yourself to adapt and recompile the boot loader. This is not an impossible task, but being a novice, you are probably not yet up-to-speed to do this successfully.

Finally, the MCU has some configuration fuses that must be programmed before you can use it. For this you also need an AVR programmer. The fuse values depend on the AVR type but, fortunately, the Arduino IDE contains all the necessary data. The IDE is capable of programming the boot loader into the microcontroller and configure its fuses for you, but your AVR programmer must be supported by the IDE.

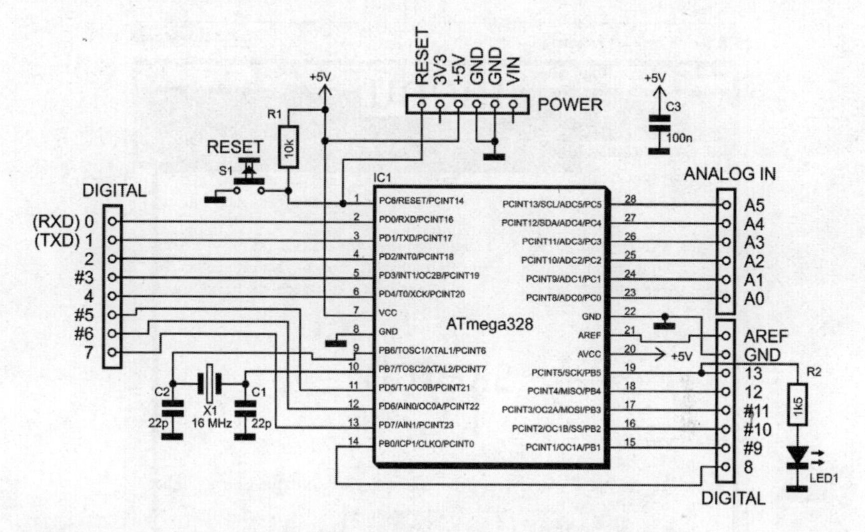

Figure 4-6 - This is all you need for an Arduino compatible board.

When you have completed your Arduino compatible board, you can give it a name. You are not allowed to call it "Arduino Something" since only boards approved by the Arduino team are entitled to the appellation. A solution is provided by the Freeduino community (`www.freeduino.org`) that allows the use of the Freeduino name for homemade Arduino boards. On their website there even is a tool to generate names that rhyme with Arduino.

4.4 The Kitchen

The details on how to install the Arduino IDE can be found in Chapter 1.

The Arduino IDE is rather simple. It is written in Java and includes:

- A rudimentary text editor to write the program;
- Some shortcuts to configure the IDE and to quickly find examples and help;
- Functions to compile the program and load it into the MCU;
- A simple terminal to communicate with the board over the serial port.

The six most frequently used functions have buttons below the main menu. From left to right we have *Verify*, *Upload*, *New*, *Open*, *Save* and *Serial Monitor*.

Figure 4-7 - The Arduino IDE and its three main areas.

A program is called a Sketch and it is written in C or C++ (or assembler if necessary).

With version 1.0.1 the IDE was extended with some major new features including menu translation into some thirty languages. By default, the IDE is in English, but it can be translated to, for instance, French, Danish or Lithuanian.

The introduction of version 1.5 of the IDE did not add many visible changes. The real novelty of this version is the addition of support for the Arduino Due board which, because it is based on a 32-bit microcontroller, needs a special programming tool chain.

4.4.1 File Menu

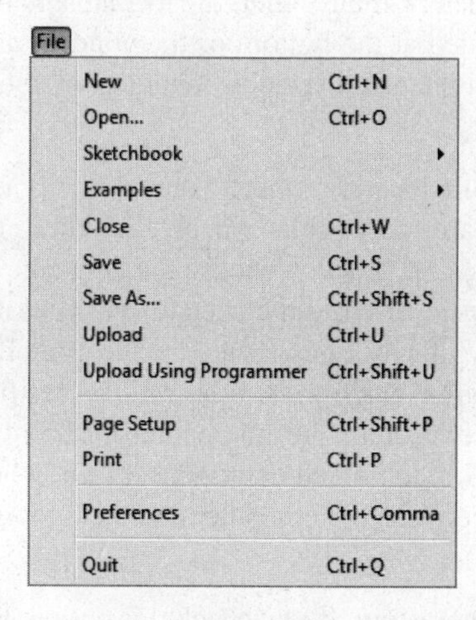

Figure 4-8 - The File menu.

New (Ctrl + N) – open a new window for a new sketch.

Open… (Ctrl + O) – open an existing sketch.

Sketchbook – open the Arduino IDE's working folder, the default location for storing sketches. Consult *Preferences* to find out where this folder is.

Examples – open the Arduino IDE's example folder.

Close (Ctrl + W) – close the active sketch.

Save (Ctrl + S) – save the active sketch.

Save as... (Ctrl + Shift + S) – saves the active sketch under a new name.

Upload (Ctrl + U) – compile and load the active sketch into the microcontroller using the serial port previously selected in the menu *Tools -> Serial Port.*

Upload Using Programmer (Ctrl + Shift + U) – compile and load the sketch into the microcontroller using the programmer previously selected in the menu *Tools -> Programmer.*

Page Setup (Ctrl + Shift + P) – prepare the active file for printing.

Print (Ctrl + P) – print the active file.

Preferences (Ctrl + comma) – display the IDE's preferences window. The window that opens shows only the most useful parameters, many others are available in the *preferences.txt* file whose location is indicated at the bottom of the window and that can be edited outside the IDE. Unfortunately, the documentation of these options leaves much to be desired.

- *Sketchbook location* specifies the default location where your sketches are saved. These are the sketches that are accessible when you click on *Sketchbook* in the *File* menu. *Editor language* lets you choose the language of the IDE. *Editor font size* sets the font used to write the sketch. By checking *Show verbose output during* the display of messages issued by the compiler and/or programmer is activated. The option *Verify code after upload* verifies that the program has been loaded correctly into the microcontroller, but it slows down the programming process. I recommend activating it only when you suspect a communication problem between the computer and the Arduino board.
- The option *Use external Editor* is also interesting. If you check this option, the IDE's editor is disabled. The IDE will compile the latest version of the sketch that you may have modified (and saved!) in an editor external to the IDE. It allows you to write the program in the editor of your choice, which is probably more powerful than the one that comes with the Arduino IDE. A disadvantage is that the sketch must be opened in two tools (the IDE and the external editor) and you have to go back and forth between the two to edit and compile the sketch.
- *Check for updates at start-up*: the Arduino IDE will contact the Arduino website to see if a new version of the IDE is available;
- *Update sketch files to new extension on save*: IDE versions before 1.0 used the file extension .PDE for sketches. With the introduction of version 1.0 of the IDE the sketch file extension was changed to .INO. Check this option to automatically change the extension of old sketches to .INO (with the result that older versions of the IDE will no longer recognize the renamed sketches).
- *Automatically associate .ino files with Arduino*: this option allows you to launch the IDE by clicking on a sketch with the extension .INO.

Quit (Ctrl + Q) – close all open windows of the IDE. Clicking on the small white cross in the red square in the upper right of a window only closes that one.

4.4.2 Edit Menu

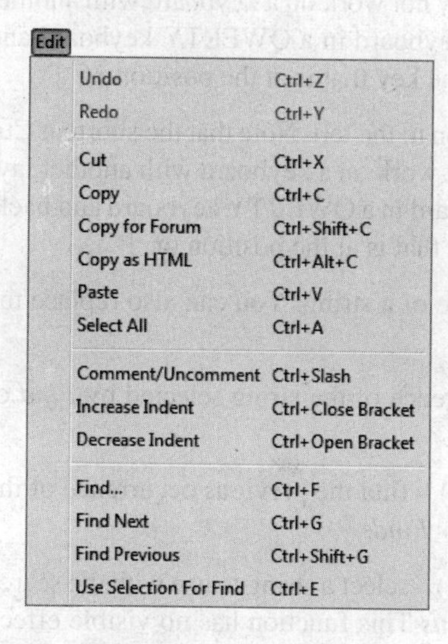

Figure 4-9 - The Edit menu.

Undo (Ctrl + Z) – cancel the last editor action.

Redo (Ctrl + Y) – redo the last editor action.

Cut (Ctrl + X) – cut the selected text.

Copy (Ctrl + C) – copy the selected text.

Copy for Forum (Ctrl + Shift + C) – copy the selected text including its colors and layout for posting it on the Arduino forum.

Copy as HTML (Ctrl + Alt + C) – copy the selected text including its colors and layout for inclusion in an HTML page.

Paste (Ctrl + V) – paste the selection that was previously cut or copied.

Select All (Ctrl + A) – select all the text in the editor.

Comment/Uncomment (Ctrl + /) – add "//" to the beginning of each line of the selected text to comment it out or remove "//" from the beginning of each line of the selection to uncomment it. Note that the shortcut Ctrl + / is for a QWERTY keyboard and may not work on a keyboard with another layout. Tip for Windows: to change your keyboard in a QWERTY keyboard and back, press Shift + Alt. Now you can find the key that is at the position of '/'.

Increase Indent (Ctrl +]) – shift the selection to the right. Note that the shortcut Ctrl +] is for a QWERTY keyboard and may not work on a keyboard with another layout. Tip for Windows: to change your keyboard in a QWERTY keyboard and back, press Shift + Alt. Now you can find the key that is at the position of ']'.

Decrease Indent (Ctrl + [) – shift the selection to the left. Note that the shortcut Ctrl + [is for a QWERTY keyboard and may not work on a keyboard with another layout. Tip for Windows: to change your keyboard in a QWERTY keyboard and back, press Shift + Alt. Now you can find the key that is at the position of '['.

Find… (Ctrl + F) – find the next occurrence of a string. You can also replace the string with another.

Find Next (Ctrl + G) – find the next occurrence of the string selected by *Find* or *Use Selection For Find.*

Find Previous (Ctrl + Shift + G, since 1.0.1) – find the previous occurrence of the string selected by *Find* or *Use Selection For Find.*

Use Selection For Find (Ctrl + E, since 1.0.1) – select a string to use with the search functions *Find, Find Next* and *Find Previous* This function has no visible effect. Select some text, for example the name of a variable, press Ctrl + E and then start the search with Ctrl + F, Ctrl + G or Ctrl + Shift + G.

4.4.3 Sketch Menu

Figure 4-10 - The Sketch menu.

Verify / Compile (Ctrl + R) – compile the sketch without loading it into the microcontroller.

Import library... (version 1.5 or later) – add a line *#include* <*xxx.h*> to the top of the sketches main file, the one that bears the name of the sketch. "*xxx*" is the name of the library to import. This option provides a sub-option Add Library... for installing libraries prepared by other users.

Show Sketch Folder (Ctrl + K) – open the folder of the active sketch in a separate window.

Figure 4-11 - The Arduino IDE does not feature debugging. For example, setting breakpoints to stop the sketch is not possible.

Add File... – add a file to the sketch. Files that can be compiled (.ino, .pde, .h, etc.) are copied into the folder of the sketch and they are opened in new tabs in the IDE. For other files a *Data* subfolder is created (if it didn't exist yet) in the folder of the sketch and the file is copied to this subfolder. Such files will not be compiled.

Import Library... (version 1.0.x) – add a line *#include <xxx.h>* to the top of the sketches main file, the one that bears the name of the sketch. "*xxx*" is the name of the library to import.

4.4.4 Tools Menu

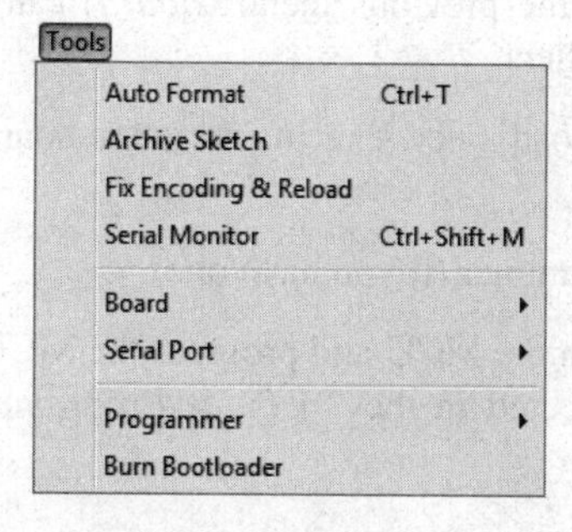

Figure 4-12 - The Tools menu.

This menu may take some time to open, especially with Windows XP. So be patient when you click on it.

Auto Format (Ctrl + T) – reposition functions to the beginning of the line but do not expect miracles. You can forget this option as soon as you have understood the importance of laying out your sketches properly.

Archive Sketch – create an archive file of the active sketch in the folder specified by *Sketchbook location* (see *Preferences* on the *File* menu). The archive will contain everything that is in the directory of the sketch. The name of the archive includes the date and a letter (a, b, c ...) to indicate the version of the day.

Fix Encoding & Reload – if (part of) a sketch contains characters that do not belong to the so-called UTF-8 character set, an error will be reported during compilation. Such characters can enter a sketch when for example the file was modified outside the Arduino IDE and saved with a different character set, such as ANSI. This function allows you to correct this problem.

Serial Monitor (Ctrl + Shift + M) – open the terminal to communicate with the board through the computer's serial port.

Board – select the Arduino board to compile the sketch for. Depending on the version of the IDE the list may be different. Starting at version 1.5 of the IDE, the boards that can be equipped with different microcontrollers only appear under their generic names and the processor must be selected from the *Processor* menu. For instance, the Mega 1280 and Mega 2560 boards are listed here under the generic name Mega. The choice of the MCU, ATmega1280 or ATmega2560, is done in the next menu. Versions 1.0.x of the IDE list all known boards.

Processor (version 1.5 or later) – select the MCU of the Arduino board. This option is active only if the board type selected in the previous menu (*Board*) can be equipped with different kinds of microcontrollers.

Serial Port – select the serial port used to upload a new executable to the board's MCU and to communicate with it.

Programmer – select the external AVR programmer (if you have one).

Burn Bootloader – program the boot loader in the MCU and program the MCU's fuses. An AVR programmer previously selected in the *Tools -> Programmer* menu is required to for this function to work.

4.4.5 Help Menu

Getting Started – open (in a browser) the Quick Start Guide.

Environment – open (in a browser) the help on the IDE.

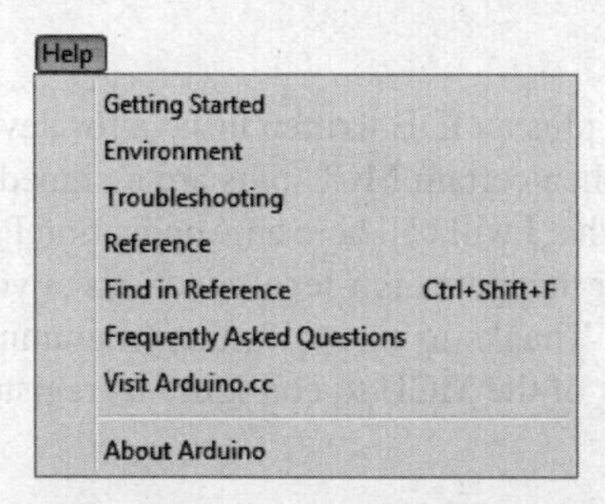

Figure 4-13 - The Help menu.

Troubleshooting – open (in a browser) a list of frequently encountered problems and their solutions.

Reference – open (in a browser) the so-called Language Reference or, more accurately, the Arduino Application Programming Interface (API) reference.

Find in Reference (Ctrl + Shift + F) – open (in a browser) the help for the function that is currently selected in the sketch.

Frequently Asked Questions – open (in a browser) the Frequently Asked Questions (FAQ).

Visit Arduino.cc – open the Arduino website in a browser (requires an internet connection).

About Arduino – display the About window that contains the credits and the IDE's version number.

4.4.6 Manage the Tabs

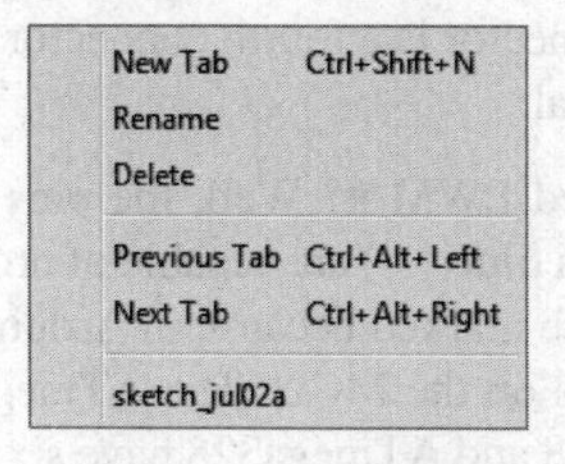

Figure 4-14 - The access to the tabs menu is not obvious.

When you open a sketch, the IDE opens all the files that (it thinks) can be compiled, found in the sketches folder and creates a tab for each file opened. On the right of the IDE, under the menu bar behind a small triangle (or down arrow, if you prefer) hides a menu with options related to tabs that allows you to, amongst others, rename, delete, and add tabs.

4.5 The Service

A board with an AVR microcontroller on it plus an IDE written in Java for developing applications only become Arduino when certain MCU pins are assigned to specific functions. This is taken care of by what I will call here an Application Programming Interface or API, a software layer that puts high-level functions at your disposal to do the low-level work for you. Thanks to this API, the programmer does not have to get down to the nitty-gritty of the MCU to configure its registers manually.

4.5.1 Table Setting

In Arduino, pin names have been simplified. The traditional notion of input-output ports (Port A, Port B, etc.) has been replaced by pin numbers. The numbering is simple and logical once you know that the first Arduino board was equipped with an ATmega8: it starts with 0 on pin 2. What do you mean, not logical? It is, as pin 1 of the ATmega8 chip corresponds to PC6, meaning I/O number 6 of port C. However, this pin is also the chip's reset input and because the designers preferred to connect it to a pushbutton, PC6 is no longer free[1]. The IC's second pin, PD0, or I/O number 0 of port D, is available and the numbering starts here. It continues counting up while skipping the pins that are not I/O pins (power supply) or that are not free for other reasons, such as PB6 and PB7 that are reserved for the crystal. In the end, we are left with 20 digital I/O pins (0 to 19). I/O 14 to 19 happen to be connected to the analog-to-digital converter (ADC) of the MCU and it was decided to group them on a separate connector and call them "analog inputs" (0 to 5). So, a Uno-type board features 14 digital I/O pins and 6 analog inputs. The 14 digital I/O pins are wired to two connectors named Digital (PWM#) on one side of the board, the 6 analog inputs are wired to a connector, called Analog, that faces the Digital connectors. In line with the Analog connector is a fourth connector named Power which also gives access to the reset signal.

Why are the digital I/O connectors also labeled PWM #? Well, the pins on the board marked with a hash ('#', sometimes with a tilde '~') are capable of providing pulse width modulated (PWM) signals. They are marked because in Arduino they serve as analog outputs. Their number depends on the MCU. The ATmega8 has only three of them (I/O 9 to 11), the ATmega168 and ATmega328 have six (I/O 3, 5, 6 and 9 to 11).

1. This is not quite correct, because by programming the fuse RSTDISBL it is still possible to use this port as a digital input connected to a pushbutton. However, doing so will break the automatic reset needed to start uploading a sketch.

For Mega-type boards that are equipped with MCUs that have more pins than the ATmega8/168/328, the numbering of the I/O pins is different, because they are grouped by function. Such a board has a Power, an Analog in, a Digital, a Communication and a PWM connector. The Digital connector even has two rows. A Mega 2560 has a total of 70 I/O pins divided over 32 digital I/O (Digital), 16 analog inputs (Analog in), 15 PWM outputs (PWM) and 10 pins reserved for Communication.

By fixing the pin functions, Arduino can do their initialization, which saves you, the programming newbie, a lot of hard work. Contrary to what you may expect, this does not limit the flexibility of the programmer, because Arduino does not impose its numbering. You are free to ignore it, do as you feel, but if you do, do not be surprised when an Arduino function does not give the desired result.

4.5.2 The Headwaiter

Strictly speaking you don't need one, but without it life is much harder. I'm talking about the boot loader. Each Arduino board comes with one pre-programmed that loads the compiled sketch – the executable – in the MCU without the need for a special programmer. An ordinary serial port is enough, although most Arduino boards have a USB port.

The boot loader is a small piece of software living in a special part of the program memory. It is always executed when the MCU restarts after an external reset (caused by the reset pushbutton for example, or by power cycling the board). The boot loader scans the serial port of the MCU for a few seconds to see if a reprogramming request arrives. If this is not the case, the boot loader launches the main application, if it exists. If, on the other hand, a reprogramming request is received, the boot loader goes into programmer mode and the IDE can send the compiled sketch to be put in memory. The whole operation takes about ten seconds; the exact time depends on the size of the executable. Once loading is complete, the board is restarted. Since this time the reset is caused internally by the boot loader itself, it will not be executed a second time. The application is launched instead.

Several boot loaders exist that differ mainly by the number of options available. However, the more options you have, the bigger the boot loader will be. The maximum amount of memory available for the sketch depends on the size of the boot loader, so if your sketch is large, having a small boot loader is an advantage. For Arduino there are boot loaders that take up less than 512 bytes and that are quite usable in most cases.

The Arduino boot loader does not use the same communication protocol as Wiring (STK500 versus STK500v2), which breaks the compatibility between the two platforms. The Wiring boot loader also takes up more space than the Arduino boot loader.

In rare cases the boot loader can be destroyed during a poorly controlled programming operation. If this happens, you can restore it from the IDE with an AVR programmer. The IDE can also program the executable directly into the MCU without using the boot loader, but again, you'll need an AVR programmer. In short, it is better to avoid overwriting the boot loader.

5. My First Offense

In computing it is a tradition to begin a programming course by writing a small program called Hello World. On a computer with a screen such a program is nothing more than a few lines of code that display the text "Hello World". On a microcontroller system without a display, one possibility is to send a welcome message over a serial port, but to blink an LED is generally simpler. Although the Arduino serial port is actually easier to use than the on-board LED, we will go for a blinking LED anyway.

As Arduino is programmed in C and C++, we need to know the syntax of these languages. C++ is a complex programming language that I cannot teach you here in a few lines. If you want to know more, I suggest you read a good book on the subject. Even though you can find every detail about C++ on the internet, a few evenings surfing is simply not enough to get to grips with this vast topic.

C++ is compatible with its simpler predecessor, C, and everything that is possible in C is also possible in C++. For what we are going to commit in this book some knowledge of C is more than enough. The necessary theory will be explained along the way, when we need it.

Figure 5-1 - The theory necessary will be explained along the way.

5.1 The Wrench

In C/C++ the main tool is the *function* (often called method in C++). We can write a single, large function that does everything, but that's not recommended. Not only does it impair the understanding of the program, but that's not practical either. We therefore cut the program up in several smaller functions that together provide the wanted functionality. Here is the function `bumming` that does nothing, but shows the basics:

```
void bumming(void)
{
}
```

A function can take arguments and return a result. The function `bumming` has no arguments and does not return a result as indicated by the two occurrences of the word **void**. The **void** on the left replaces the result to return, the **void** on the right replaces the arguments. I wrote **void** in boldface because it is a reserved word or keyword of C that cannot be used as a function name, a variable name or as the name of a constant.

C and C++ are case sensitive, meaning that uppercase and lowercase characters are not interchangeable. `Void` and **void** therefore do not mean the same thing and the compiler will interpret `Void`, `voiD`, `VoId` and `VOID` as names of different functions, variables or constants.

The arguments, present or not, are always listed between two parentheses "()". Commas separate the arguments when there are more than one. If there are no arguments, it is allowed to omit the keyword **void**, the parentheses may remain empty, but I do not recommend this practice for safety reasons. In fact, writing explicitly the keyword **void** tells the compiler – and the reviewer of your code – that the arguments have been left out intentionally. The **void** which replaces the return value (see below) cannot be omitted[1].

```
float add(int apples, int pears)
{
  return NAN;
}
```

1. Strictly speaking this is not true. Most compilers will suppose that a function of which the return data type is not specified will in fact return an integer (**int**).

The arguments are preceded by the type of data they contain. In fact, it are the data types that really matter, the names of the arguments can be omitted. Data types are discussed later in Section 5.4 (page 62). The example `add` above expects two integers as arguments. The result is a floating-point number. `NAN` means Not a Number and is a C constant used to indicate that a calculation is impossible (you cannot add apples to pears).

Braces "{}" surround the body of the function, i.e. all instructions executed by it. Semicolons ';' separate the individual instructions. To improve readability it is preferable, but not required, to write only one statement per line. Basically, an instruction may take the form of a call of another function, of a calculation or of a control structure such as a loop or a condition.

The reserved word **`return`** is used to return the result. The returned value may be obtained from a function or a calculation, but a variable or constant also can do. If the **`return`** statement is missing in a function that is not **`void`**, the compiler will generate a warning or an error[1]. The **`return`** statement may be placed almost anywhere in a function and it is even possible to put it in several different places. On the other hand, it is mandatory to put a **`return`** at the end of a non-void function, or more precisely, it must be impossible to leave the function without passing a **`return`** statement. A void function may contain a **`return`** statement without a return value. This may be used to leave a function before the end:

```
void honey_i_am_gone(int wallet)
{
  if (forgot_credit_card==TRUE) return;
  else off_to_vegas(wallet);
}
```

Here's my reasoning: if I forgot my credit card, I will return home, otherwise I will continue my trip to Las Vegas. The variable `wallet` is not used here (because I do not keep my credit card in my wallet I can detect its absence without touching my wallet), but I'm going to need it at the destination. The conditional structure **`if-else`** in this example is discussed later in Section 5.6 (page 64).

This example shows something else: whitespace in the names of functions, variables and constants are prohibited. Several ways to attach the words that make up a name are in vogue; personally I always replace the space with the underscore '_'.

1. A compile error must be corrected in order to continue, a warning may be left as is. However, anyone as serious about programming as you will always try to correct all warnings.

5.2 Get to Know the Hood

The Arduino program (or sketch) in its simplest form consists of only two functions: `setup` and `loop`[1]. They take no arguments and do not return a result. The first is executed only once, at the start of the application and everything related to its initialization goes in here. It is in `setup` that you define which pins will be inputs, which pins can control something, at what rate the serial port will communicate, etc.

Since many programs execute an endless loop, repeating the same operations over and over again, Arduino sketches contain the `loop` function that will be at the heart of your application.

It is convenient to create additional functions to improve the readability of the program when it becomes more complex. You can add as many functions as you want, as long as the compiled sketch will fit in the MCU's program memory.

Both functions **`setup`** and `loop`, empty or not, must be present in every sketch. If you forget one of them, or both, the IDE will generate an error message of the type undefined reference. This error that often makes beginners panic says that somewhere in the program you refer to something (a function, a variable or a constant) that the IDE (the linker actually, see also Section 3.3.1, page 32) cannot find. The IDE hides much of the complexity of Arduino, but it does not protect you against the sometimes incomprehensible messages generated by the tool chain that it controls. The IDE is like a boat on a calm blue ocean without any apparent danger but from which at all times cryptic messages may arise from the depths like sharks. Do not panic, stay calm and look for the error – usually a typo – in your code.

The function `loop` is like any other function even though it may seem special. Just like the function `setup` it is called from a lower layer (the `main` to be exact) and its local variables (i.e. the variables declared inside the function) are recreated on every call. You must declare variables outside the functions to make them keep their values between calls of `loop` (such variables are called global).

The function `loop` is not called periodically but continuously, meaning that as soon as it is finished, it is called again. The calling frequency depends on the time required to execute the function. If `loop` or another function called from `loop`, blocks, the entire application, including `loop`, will be blocked. There is no magic mechanism that keeps calling `loop`.

1. Outside Arduino a C or C++ program usually has to contain at least the function `main`.

For your information, the minimum execution time of an empty `loop` function is 750 ns, when compiled with Arduino 1.0 for an Arduino Uno board equipped with an ATmega328P clocked at 16 MHz (12 machine cycles, GCC version 4.3.2 (WinAVR 20081205)).

5.3 Preparing the Job

In what follows, you will need an empty sketch. Launch the IDE or open a new sketch. Unfortunately, the Arduino IDE does not provide the functions `setup` and `loop` when you create a new sketch. Even if they are required, they have to be added manually. Here the Arduino team clearly missed an opportunity to create a better world.

The function `setup` of our sketch has not much to do. We need an output for the LED and that's all. Since all pins are inputs by default, we will transform one into an output. To do this we must call the function `pinMode` with as arguments the number of the pin and its mode `INPUT` or `OUTPUT`. The third mode, `INPUT_PULLUP`, which was introduced in version 1.0.1 of the Arduino software, will be discussed in the next chapter. The LED on the board is connected to pin 13, so we can write:

```
void setup(void)
{
  pinMode(13,OUTPUT);
}
```

We could leave it at this, but it is not practical for the rest of the sketch. Indeed, in the loop (the function `loop`) we will also have to specify the pin that is connected to the LED. To reduce the risk of errors, here is a programming trick: it is better to define a constant variable to store the number of the pin. This allows the compiler to check if there are no mistakes, something it cannot do with plain numbers. For example, the compiler will recognize and process `led` and `lad` as two separate variables that must be declared explicitly. If you use both, but declared only one, the compiler will generate an error. Numbers on the other hand do not need a declaration prior to being used and 13 and 14 for instance look the same to the compiler, they look like numbers. The compiler does not notice if you mixed up numbers and it will not report an error.

Summarising, we define a variable named `led` and we give it the value 13:

```
int led = 13;                    // the LED is connected to pin 13

void setup(void)
{
  pinMode(led,OUTPUT);           // the LED is connected to an output
}
```

The variable led declared outside of the function is a global variable. Unlike a local variable which is declared inside a function, a global variable is available (programmers say "visible") for all functions defined after the variable's declaration. Here the variable must be global, because both setup and loop need it. We will see on page 65 an example of a global variable and how it is used.

5.4 A Sizeable Problem

In the code fragment above, the word **int** means integer, therefore the variable led is an integer and it has the value 13. You can use this variable anywhere where the C language accepts integers. There exist more predefined data types, such as the **char** (character), the **float** (a floating-point number, i.e. a number with decimals) and the **double** (a floating-point number with better accuracy than a **float**, i.e. with more decimals). It is also possible to define your own data types.

The size in bytes of a data type determines the range of values that such a variable can contain. In Arduino a **char** is one byte, an **int** is two bytes, a **float** is four bytes and a **double**... also four! Usually a **double** occupies twice as many bytes as a **float**, hence the name, but not in Arduino. Here they do not increase the accuracy of calculations.

This illustrates a problem of the base data types in C: their sizes in bytes are not defined by the standard, but by the platform. To overcome this difficulty in microcontroller programming, it is customary to define data types more accurately and you often encounter data types like byte (8 bits) and word (16 bits). Type definitions that include the size in bits like int8 or uint32 are even more explicit. The "u" in uint32 indicates that this is an **unsigned** integer data type, meaning that it is always positive or zero. Defining data types in this way allows you to recompile the same source code for a different platform without too many surprises at this level (other surprises at other levels are not to be excluded).

Additional data types have been defined in Arduino, like the boolean and the string (a string of characters, not the garment).

5.5 The Snitch

Machines do not care, but for humans comments are crucial if you want your program to be understandable for other people. When you are actively developing a program you will necessarily remember the purpose of each function, but when you look again at the same program a few months or years later, it is likely that you have forgotten some of the program's subtleties. Comments were invented to help the person who reads the code to understand what all the functions are supposed to do and how. Comments can document a program.

In C, comments are surrounded by "/*" and "*/", like this:

```
/* Magritte would have said: this is not a comment */
```

Everything between /* and */ is ignored by the compiler, even if the comment spans multiple lines, or entire pages. The character sequence /* and */ may seem a bit strange, but when you look at the numerical keypad on your PC keyboard (if it has one) it all becomes clear. Indeed, these keys are placed next to each other and this sequence was chosen for ergonomic reasons (or out of laziness).

In the function `setup` that we saw before there were also so-called end-of-line comments recognisable by the sequence "//" that precedes them. Whatever comes after // on the same line is ignored by the compiler. Not all C compilers know about end-of-line comments as it is one of the few C++ features that have been adopted by the C standard (by the C99 standard to be exact) and only the compilers that conform to this standard will (and must) accept end-of-line comments.

Be careful to avoid exotic symbols like accented characters in the program and in comments. When your mother tongue is English, this will not often be a problem, but some languages use many accents and special characters like ¿, ç, å or ©. Such characters must be avoided, because some compilers have problems digesting exotic characters, even when they only appear in the comments where they should be ignored. In C, it is best to use only the basic English alphabet (a-z and A-Z); accents and other special or exotic characters are excluded.

Be careful when end-of-line comments are mixed with "slash-star" comments or when a piece of code that already contains comments is commented out, because it can disrupt the compiler. If you get an error or a warning that talks about nested comments check that you did not forget a */ somewhere. Do not rely on the syntax coloring of the IDE, an editor is not a compiler!

5.6 Flashing Lights

Now that we have a variable `led` and the board's LED connected to a pin that is configured as an output, we can move on to the function `loop`. To turn on and off the LED, the Arduino API provides the function `digitalWrite`. It takes two arguments: the number of the pin and the value `HIGH` (a high logic level) or `LOW` (a logic low). If we call this function first with the value `HIGH` and then with the value `LOW` and if we wait a bit between the two calls, the LED will flash as this sequence is repeated continuously because of the function `loop`. The API function `delay` is available to create delays and using it will make the program wait. The time to wait is specified in milliseconds.

Here is the finished sketch:

```
/*
 * blink 1
 */

int led = 13;

void setup(void)
{
  pinMode(led,OUTPUT);
}

void loop(void)
{
  digitalWrite(led,HIGH);       // switch on the LED
  delay(500);                   // wait 500 ms
  digitalWrite(led,LOW);        // switch off the LED
  delay(500);                   // wait 500 ms
}
```

This is a way of doing things, but it is not the only way. This sketch allows you to specify separately the LED's on and off time. If this is not necessary, you can modify the sketch to specify a single on and off time, like this:

```
/*
 * blink 2
 */

int led = 13;
int state = HIGH;

void setup(void)
{
```

```
  pinMode(led,OUTPUT);
}

void loop(void)
{
  digitalWrite(led,state);      // switch the LED on or off
  if (state==HIGH)
  {
    state = LOW;                // toggle the value of state
  }
  else
  {
    state = HIGH;
  }
  delay(500);                   // wait 500 ms
}
```

At the beginning of `loop` the LED is switched on or off, depending on the value of the variable `state`. If this variable is equal ("==", two times '=', not equal is indicated by "!=") to `HIGH`, we will give it the value `LOW`. If not (**`else`**) we will give it the value HIGH. This way the value of `state` is inverted every time the program executes `loop`.

Note how the variable `state` is declared outside the functions. Like the variable `led`, `state` is a global variable. If I had declared it inside the function `loop`, it would have lost its value every time the function was exited and the sketch would not have worked as intended.

5.7 Incarcerated

Now that we have a sketch with a function `setup` and a function `loop`, it is time to try it out. For this it is necessary that your Arduino board is connected to your computer and that the driver for the board is installed. If you haven't done this yet, refer to Section 1.2 (page 11) and follow the instructions.

- In the menu of the IDE click on *Tools*, then on *Board* and select your board from the list that appears. If the version of your IDE is 1.5 or later, it may be that you also have to choose a processor from the *Processor* submenu. What to choose depends on your board, so be careful with this option. If it is not greyed out, you must choose a microcontroller or at least verify that the microcontroller selected by default corresponds to the one on your board.

- In the menu, click on *Tools* followed by *Serial Port* and select the serial port to use. If you do not know the name of the port, refer to Section 1.3 (page 16).

✦ Click on the round button with the arrow to the right (or *Upload*) to compile the sketch and load the executable file into the MCU. If you are using an old Arduino board, it may be necessary to reset the board manually before you can upload a program, but normally this is done automatically.

Once the sketch is uploaded to the MCU the LED 'L' on the board, the one that is connected to digital output 13, should start flashing at a frequency of 0.5 Hz.
Did it work? Congratulations! Now quickly move on to the next Section.
It did not work? Then read Section 5.9 (page 70); it may help you solve your problem.

5.8 Out on Parole

The last sketch, `blink 2`, introduced an important element of the C language, the structure **`if-else`**, a conditional structure. The condition is specified between round brackets ("()") following the **`if`**, and the instructions written between braces ("{}") following the condition are executed if the condition evaluates to true. Otherwise the part between the braces following the **`else`** statement will be executed.

```
if (condition)
{
  ...
}
else
{
  ...
}
```

The parts between and including the braces are optional, as is the **`else`** part, but **`else`** cannot be used without **`if`**. If the braces are missing the instructions placed right after the **`if`** or **`else`** keyword or on the next line will be executed. Here are some examples (remember, "==" means "equal to"):

```
if (you_like_to_live_dangerously==TRUE)
{
  be_careful();
}
else
  be_careful_anyway();
  you_can_die_in_traffic_too();
```

This example shows the danger of omitting the braces after **else**. The second line following the **else** statement is always executed as it is not part of the conditional structure even though the indentation suggests otherwise. Only the execution of the first line following **else** depends on the state of the condition. The same applies to the following example. The instruction on the line following **if** is always executed, but the danger is less, because the indentation does not suggest something else.

```
if (your_spouse_is_a_good_cook==FALSE) order_a_pizza();
enjoy_her_company();
```

When there is only one statement following **if** or **else** then it is better to write it on the same line. The best solution however is to always use braces to improve the readability and understanding of your program, but also to avoid preventable errors:

```
if (your_dog_is_mean==TRUE)
{
  notify_the_postman();
}
```

We can do the same thing with an empty **if** and an **else** with body by inverting the condition, but it is not very intuitive:

```
if (the_postman_is_friendly==TRUE) ;
else
{
  buy_a_mean_dog();
}
```

The C language offers a second method for handling conditions, the **switch-case-default** structure:

```
switch (variable)
{
  case x:
    break;

  case y:
    break;

  default:
}
```

The reserved word **switch** acts as a kind of logical switch that connects the value of variable to several conditions, all indicated by the reserved word **case**. When there is no **case** that evaluates to true, the program will execute the instructions following the keyword **default**. This keyword is optional, like the **else**. You can put as many conditions as you want, as long as there are no duplicates. This is why the **switch-case** structure is often encountered in state machines that have many states. If there is only one condition, the **switch-case-default** is identical to the **if-else** structure, writing

```
switch (variable)
{
  case x: …;
}
```

```
is the same as
```

```
if (variable==x)…;
```

Do not forget the braces in a **switch-case** structure, they are mandatory.

Unlike an **if-else** structure that requires that the specified condition is true or false, the **switch** variable must contain an integer, or, more precisely, something that the compiler can interpret as an integer, because it must compare the contents of this variable to the numerical values following the **case** statements. The **case** may not be followed by a floating-point value or a character string[1].

The conditions must be equalities, intervals like “smaller than” or “bigger than” are not allowed. Intervals can be done using several conditions, as follows:

```
switch (blood_alcohol_content_as_percentage)
{
  case 0:
    have_a_drink();
    break;

  case 3:
    entering_the_pleasure_zone();
    break;

  case 10:
    why_don_t_you_lie_down_a_bit();
    break;
}
```

1. A character **char** is considered to be an integer value.

Conditions may appear in any order and there may be gaps as in this example where there isn't anything between 0 and 3 and 10. The same functionality with an **if-else** structure looks like this:

```
if (blood_alcohol_content_as_percentage==0)
{
  have_a_drink();
}
else if (blood_alcohol_content_as_percentage==3)
{
  entering_the_pleasure_zone();
}
else if (blood_alcohol_content_as_percentage==10)
{
  why_don_t_you_lie_down_a_bit();
}
```

If you forget the **break** at the end of a **case**, the program continues with the instructions following the next **case** (or **default**) without testing the validity of this **case**. Execution continues until a **break** or the end of the structure is encountered. It is therefore possible to do this:

```
switch (number_of_kids)
{
  case 3:
    almost_large_family();

  case 1:
    almost_ordinary_family();
    break;

  case 2:
    ordinary_family();
  break;

  case 0:
  default:
    extraordinary_family();
}
```

According to this example, a family with three children is almost a large family and almost ordinary at the same time. The last **case** is useless. Why?

5.9 Reintegration

If you did not manage to get the sketch presented in this chapter to work, you have a problem. Let's try to find the solution together. For starters, is the power LED of the board on? No? Check that the board is powered. As a matter of fact, does your board possess such an LED?

Some boards have the option to disconnect the power supply from the USB port. Is there a small switch or a jumper near the USB connector or near the external power supply connector? Get your multimeter out and try to find traces of the supply voltage on the board. Do you have 5 V on the MCUs VCC pin? (pin 7 on a 28-pin AVR)

If the board is powered on (the LED is on), verify that it is recognized by your computer. Read Section 1.2 (page 11) to find out how to install your board properly. If everything works correctly, check that you did not make a mistake while selecting your board and serial port in the IDE menus, as was described in Section 5.7 (page 65).

If all this is fine, look closely to what happens in the IDE when you click on the *Upload* button. Are there any error or warning messages? If so, try to correct your sketch. If not, check that the sketch uses indeed pin 13 to blink the LED. Finally, if you really could not find any errors, check that there is an LED connected to pin 13. Also, do not forget that an LED has a "polarity", perhaps it is connected the wrong way around? BTW, LEDs can break too. Are you sure the LED is OK?

6. Digital Signals: All or Nothing

Of all the peripherals available on a microcontroller, the digital inputs and outputs are the most common and the most widely used. In the technical literature they are called General Purpose Input Output or GPIO ports. Configured as outputs, these pins are mainly used to provide signals which do not change state often, to command transistors and LEDs, to enable or disable other ICs or to power sub-circuits. Configured as inputs they allow reading the state of pushbuttons and switches, they can receive interrupt request signals, etc. Although in many applications these signals are rather slow, this does not mean that the digital I/O are not able to produce or follow fast signals. On the contrary! Digital outputs can for instance be used to deliver fast clock signals for other chips. Digital inputs and outputs can also be combined to create custom serial or parallel communication interfaces or, on the other hand, create standard serial or I²C communication interfaces if the MCU you are using is lacking one. In this case, we talk about soft(ware) communication ports to distinguish them from communication ports implemented in hardware.

6.1 Surprises

In Arduino the MCU's digital I/O lines are easy to use with only three functions: `pinMode`, `digitalRead` and `digitalWrite`. With `pinMode`, a pin is configured as an input or output; the state of an output is controlled by `digitalWrite` and an input can be read with `digitalRead`. In the previous chapter we have seen how to set up and use an output, and using an input is very simple too. Here is an example of a short sketch full of surprises. It reads the value of an input to decide whether or not to light the LED of the Arduino board.

```
/*
 * key 1
 */

int led = 13;
int key = 8;

void setup(void)
{
  pinMode(led,OUTPUT);
}
```

```
void loop(void)
{
  int val = digitalRead(key);
  digitalWrite(led,val);
}
```

The first surprise is the input that I have named `key`. Why? Because it makes sense to connect a pushbutton or a keyboard key to this input to control the LED. At this point I would also like to surprise you with a statement about language: please get used to programming in English. This may seem like a strange statement in a book written in English, but since this book will (unfortunately) not be translated in every possible language, chances are that your mother tongue is not English and you may be tempted to use your native language for naming variables and functions. However, since in Arduino everything is based on openness, sharing and mutual aid, a common language is needed to share sketches and libraries with users around the world. This common language is English. That said, if you still prefer to program in French, German or Zulu, please do so.

The third surprise is the absence of the instruction that configures `key` pin as an input. The fourth surprise? Do not connect a pushbutton.

Let's start with the third surprise, the missing statement `pinMode(key,INPUT)`. As I already mentioned in Section 5.3 (page 61), by default, at the beginning of a sketch all pins are configured as inputs and it is therefore not really necessary to reconfigure the `key` pin as an input. In fact, this is probably true for all pins able to function as input and output of all microcontrollers of the world. To be more precise, after a reset, microcontrollers tend to disconnect their outputs and configure their pins that can be inputs as inputs. The goal of this, say, survival instinct is to avoid accidentally switching on a motor or another actuator, which could possibly cause material damage or, worse, personal injury. The fact that Arduino respects the all-pins-are-inputs paradigm is therefore not so much a feature, but simply the normal state of the Arduino MCU after a restart. However, even if we can, it is not smart to omit the `pinMode(key,INPUT)` statement, because it creates uncertainty about the function of the key pin elsewhere in the sketch. In programming, it is good practice to eliminate all uncertainties about the way a program is supposed to work in order to avoid bugs and unexpected behavior. Yet another good habit for you to adopt.

Let's return to our pushbutton, why did I tell you not to connect it? Because I wanted to highlight a subtlety of the MCU. After you programmed the sketch into the MCU, what did the LED do? It may have come on, it may have stayed off or maybe it started blinking randomly? You can influence the state of the LED by touching the `key` pin with your finger. As you can see, the MCU's inputs are very sensitive to disturbances such as the voltage present on your skin. This is another form of uncertainty in the system: when an input is not connected, when it is left floating, its state is undefined. This kind of uncertainty is often removed with a resistor that forces the input to a well-defined voltage level, such as 0 V (pull-down) or the supply voltage (pull-up). These resistors are so common in microcontroller systems that most MCU manufacturers integrate them in their products. The AVR on the Arduino board also has them. On power-up these resistors are disabled, again for safety reasons, to avoid potential conflicts with the electrical circuitry connected to a pin that is configured as an input.

Since Arduino version 1.0.1 the function `pinMode` has a new mode, `INPUT_PULLUP`, which allows you to configure a port as an input with the pull-up resistor enabled. The mode `INPUT` now disables the pull-up resistor[1]. In the versions of Arduino older than 1.0.1, you had to *write* an input to enable or disable the pull-up resistor, which is not very intuitive. This technique that still works for compatibility with sketches written for an older Arduino version is due to the AVR, not to Arduino, and shows once more why it is important to explicitly configure an input pin. If you do not do it, you risk taking it for an output:

```
/*
 * key 2
 */

int led = 13;
int key = 8;

void setup(void)
{
  pinMode(led,OUTPUT);
  pinMode(key,INPUT_PULLUP);
}

void loop(void)
{
  int val = digitalRead(key);
  digitalWrite(led,val);
}
```

By modifying the sketch this way two doubts are cleared: `key` is now clearly an input, and its default state is `HIGH`. Setting its default state to `LOW` is not possible, because the AVR does not have pull-down resistors, but you can always connect an external resistor between the input pin and 0 V in case you need one.

1. Which it did not do in the versions previous to 1.0.1! Sketches that rely on the old behavior may stop working when they are recompiled with Arduino 1.0.1 or newer.

6.2 More Surprises

After the surprises of the sketch described in the previous paragraph, there is yet another one. On my computer, using Arduino 1.0.1, the compiled sketch occupies 1058 bytes. Now combine the two lines of the function `loop` into one, like this:

```
/*
 * key 3
 */

int led = 13;
int key = 8;

void setup(void)
{
  pinMode(led,OUTPUT);
  pinMode(key,INPUT_PULLUP);
}

void loop(void)
{
  digitalWrite(led,digitalRead(key));
}
```

Look at the call of `digitalWrite` in the function `loop`. Its second argument is a function call instead of an integer. This may seem strange, but in C it is perfectly legal to pass a function call as an argument to another function as long as the result value of this call is consistent with the argument. In clear: since the result of `digitalRead` is an integer, we can use it in place of any other integer, like we did here for the second argument of `digitalWrite`.

On my computer, compiling this version of the sketch creates an executable of 1064 bytes, which is 6 bytes more for exactly the same functionality as sketch `Key 2`. Strange, isn't it? It is. Here we touch on the internal processes of the compiler that are not obvious to understand. Experimenting a bit with the code, we can determine more accurately the origin of this size difference. It seems due to the use of the three variables `led`, `key` and `val` that are of type **`int`**. In the second version of the sketch the variable `val` does not appear at all and if we replace all instances of `key` and `led` by their respective values, the size difference disappears. There is no need to dwell on this oddity, but it shows that in theory harmless differences can have sizeable consequences.

In the first version of the sketch, it is as if the programmer is trying to help the compiler by adding the variable `val` that besides perhaps a bit of readability adds nothing to the sketch. In the second version of the sketch, the compiler must figure it out all by itself. But, for the compiler, in both cases it is simply C and it doesn't need your help. As a general rule, never try to help the compiler or try to be smarter than it, it is useless, as the compiler will always have the last word. That said, there are many cases where, as here, a reorganization of the code can influence the size of the executable and/or its execution speed. The final result depends much on the design of the compiler and MCU, but it is good to know about this when you have to scrape a few bytes of the executable when the sketch turns out to be too large to fit in the MCU's program memory.

Now you can connect a pushbutton between pin 8 and ground (GND or 0 V) and play with the sketch. Have fun!

6.3 The Matrix Keyboard

In the previous section, the inputs and outputs were configured at the beginning of the sketch and they kept the same role for the rest of the sketch. This is not mandatory, and it is quite possible to dynamically reconfigure a pin to change function. In some cases this is a good technique to save I/O ports, in other cases it may even be necessary in order to get the job done.

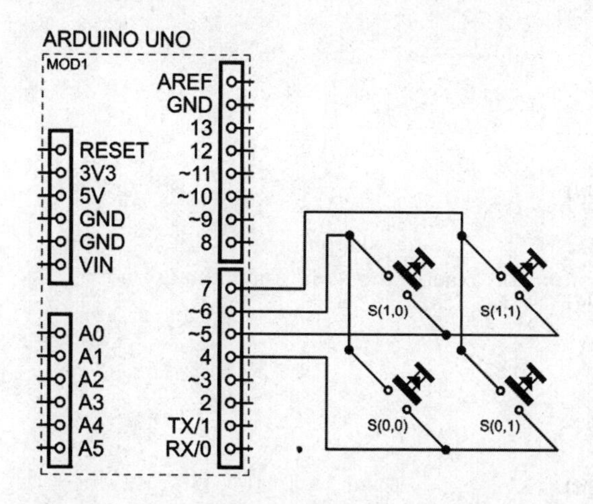

Figure 6-1 - The only advantage of a four-key matrix keyboard is its simplicity which allows you to easily understand the way it works.

Consider a keyboard. To make a keyboard, if you use the pull-up resistors of the AVR, the easiest way is to connect the pushbuttons between the MCU's inputs and 0 V. Up to three keys this technique is the most effective considering the number of I/O ports needed. For four or five keys there is no difference between this technique and a matrix of two by two, but beyond five keys, a matrix keyboard needs less input and output lines. Multiplexing is used to read such a keyboard, meaning that the keys are scanned one by one. In theory this technique can miss from time to time a short key press, but the risk is low if the multiplexing frequency is high enough. Here's how you can make a driver for a 2 × 2 matrix keyboard or keypad:

```
/*
 * key matrix 1
 */

int row0 = 4;
int row1 = 5;
int col0 = 6;
int col1 = 7;


void setup(void)
{
  Serial.begin(115200);
  pinMode(col0,OUTPUT);
  pinMode(col1,OUTPUT);
  pinMode(row0,INPUT_PULLUP);
  pinMode(row1,INPUT_PULLUP);
  digitalWrite(col0,HIGH);
  digitalWrite(col1,HIGH);
}

void loop(void)
{
  digitalWrite(col0,LOW);
  digitalWrite(col1,HIGH);
  if (digitalRead(row0)==LOW)
  {
    Serial.println("S(0,0)");
  }
  if (digitalRead(row1)==LOW)
  {
    Serial.println("S(1,0)");
  }

  digitalWrite(col0,HIGH);
  digitalWrite(col1,LOW);
  if (digitalRead(row0)==LOW)
  {
    Serial.println("S(0,1)");
  }
  if (digitalRead(row1)==LOW)
  {
    Serial.println("S(1,1)");
  }
```

```
  delay(250);
}
```

In this sketch we use for the first time the Arduino serial port, available on all boards. The Arduino Mega even has four of them. The IDE contains a simple serial port monitor that allows you to send characters to the serial port and receive those sent by the board. Click in the IDE on the top right button named Serial Monitor to open it.

In the function `setup`, using `Serial.begin`, the serial port is set to a data rate of 115,200 baud. The matrix keyboard's inputs and outputs are defined and the pull-up resistors for the inputs are activated. The columns are deactivated by a high logic level and the matrix is at rest. The pin numbers were chosen at random, but do not use pins 0 and 1 because they are needed by the serial port.

In the function `loop`, first column 0 – `col0`, in C counting almost always starts at zero – is activated (active low) and column 1 (`col1`) is disabled. The logic level on `row0` is read and if it is equal to `LOW`, then key S(0,0) must be closed and the function `Serial.println` will send the text string "`S(0,0)`" terminated with a newline character (indicated by `ln` at the end of `println`) to the computer. If, on the other hand, `row0` is read as `HIGH` due to its pull-up resistor, then key S(0,0) is considered inactive and the sketch does not send anything to the serial port. The program applies the same procedure to `row1` and key S(1,0).

Then column 0 is disabled, column 1 is activated and the two rows are read again, this time to determine the status of keys S(0,1) and S(1,1).

At the end of `loop`, the function `delay` slows down the skecth to 4 Hz to prevent the program of flooding the serial port with key status messages when you press a key. In real life, in order not to miss key events, such a loop is usually run at a frequency between 100 Hz and 1 kHz.

6.4 Charlie to the Rescue

The previous sketch used four I/O ports to read the status of up to four keys. However, by adding some diodes it is possible to do the same thing with only three I/O ports. In this case we speak of charlieplexing, a multiplexing technique named after its inventor Charlie Allen, who, at the time of invention, worked for chip maker Maxim[1]. Unlike conventional multiplexing, in charlieplexing the ports of

1. According to the internet, our trustworthy source of not always reliable information that is often difficult to verify.

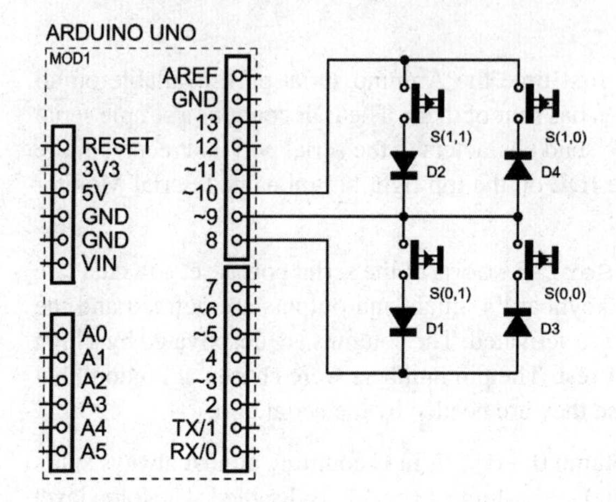

Figure 6-2 - Charlieplexing can scan four keys with only three I/O ports. The price to pay is four diodes.

the MCU continually change role. One moment they are configured as inputs, the next they may be configured as outputs, which means that they must be reconfigured on the fly. The following sketch shows how this works.

```
/*
 * Charlieplexing
 */

int row0 = 8;
int row1 = 9;
int row2 = 10;

int key_read(int r0, int r1, int r2)
{
  pinMode(r0,OUTPUT);
  digitalWrite(r0,LOW);
  pinMode(r1,INPUT_PULLUP);
  pinMode(r2,INPUT);
  return digitalRead(r1);
}

void setup(void)
{
  Serial.begin(115200);
  pinMode(row0,INPUT);
  pinMode(row1,INPUT);
  pinMode(row2,INPUT);
}
```

```
void loop(void)
{
  if (key_read(row1,row0,row2)==LOW)
  {
    Serial.println("S(0,0) active");
  }

  if (key_read(row2,row1,row0)==LOW)
  {
    Serial.println("S(1,0) active");
  }

  if (key_read(row0,row1,row2)==LOW)
  {
    Serial.println("S(0,1) active");
  }

  if (key_read(row1,row2,row0)==LOW)
  {
    Serial.println("S(1,1) active");
  }

  delay(250);
}
```

Already mentioned in Section 5.1 (page 58), but not illustrated up to now, this sketch finally shows a function, `key_read`, that has arguments, integers in this case, and that returns a result, also an integer. As a matter of fact, this function returns the result of another function, `digitalRead`. Thanks to the function `key_read` I was able to keep this sketch short, which greatly improved its readability. Another good and certainly not the least important reason to use a function is the reduction in the number of places where you can make a mistake. As you can see, the function `key_read` is called four times in the function `loop`, which gives you four opportunities to make a mistake. If, instead of using a function, I had inserted four times the five statements of the function, I would have had five times more lines of code to test. Generally, when you find yourself copying a few lines of code, ask yourself the question if it would not be better to create a function instead.

The three MCU ports are configured in `key_read`, not in `setup`. In `setup`, they were only, say, disabled. The function `loop` calls the function `key_read` four times, meaning that on every pass through `loop` the ports are reconfigured four times. The ports do not have a well-defined direction; they are a bit like quantum particles that can be in several states at once.

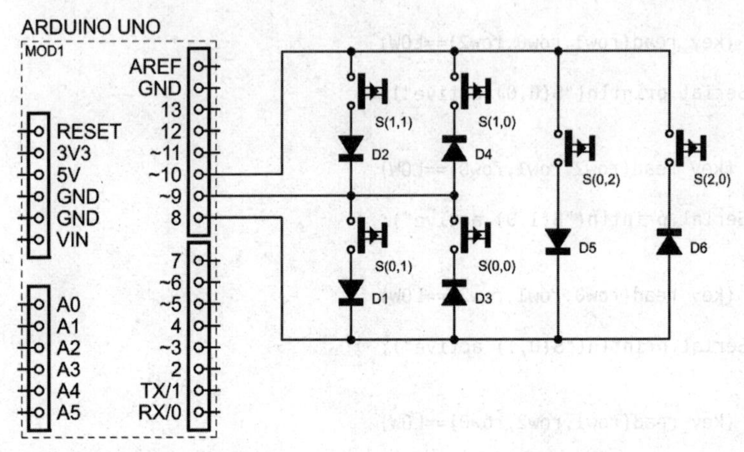

Figure 6-3 - Charlieplexing: six keys on three I/O ports for the price of six diodes.

The function `key_read` has three arguments that commute four times to scan the four keys. However, three arguments allow for six permutations, meaning that there are two unused possibilities. Indeed, charlieplexing can handle six keys with only three I/O ports; doing the same thing with a conventional matrix keyboard would require five I/O ports.

Charlieplexing also has some disadvantages, including the impossibility to avoid so-called phantom keys. This phenomenon occurs when you press multiple keys at once. In this case, we cannot determine with certainty which of the keys are pressed. A related problem are hidden keys, keys that become "invisible" when multiple keys are held down at the same time. Traditional multiplexing can avoid these complications by using a diode in series with each key (see Section 6.8 below).

A fun variation on the charlieplexing theme is LED multiplexing: replace the keys by wire bridges and the diodes with LEDs. Now it is no longer necessary to reconfigure the ports, just make them all outputs and play with their logic levels to turn some of the LEDs on or off.

Figure 6-4 - Mozartplexing: 10 pins control 88 keys.

6.5 Repeat Yourself

Many applications such as calculators, synthesizers or telephones are easier to use when they have more than four or six buttons. Fortunately, it is not difficult to extend a matrix keyboard, simply add rows and columns. To read each key, you must also add a few lines of code. For the charlieplexing sketch above we created a function to read a key that had to be called for every key. However, only the function arguments changed for each call and they always changed in the same way. The sketch would have been shorter and clearer if we had used a loop to scan the keys. In the loop the function would have been called and its arguments would have been determined. So, how do we do loops?

The C language provides three types of loops: **`for`**, **`while`** and **`do-while`**. This is their story.

6.6 The Story of the Three Loops

On a beautiful summer evening, while her family is sprawled in front of the TV, Goldilocks decides to write a small program that can add five times two variables. In her heavy C language reference book "C for Bears", she finds three techniques for introducing a loop in a program. Listed in alphabetical order, she decides to try her luck with the first one that listens to the sweet name of **`do-while`**. Goldilocks starts to code, but she is quickly lost in counting. Should I execute the loop four or five times? wonders Goldilocks, a little disturbed. Pff, she sighs, already discouraged: Perhaps without the **`do`** the **`while`** loop will be easier? She starts programming again, but cannot make the **while** loop stop in time. Oh My Gosh! (or words of similar meaning) Goldilocks cries out: This is, like, so complicated! In tears she finally turns to the **for** loop, and, surprise, she gets the loop right straight away. The **for** loop is neither too disturbing neither too complicated, it is just right for Goldilocks problem.

From the previous you will have understood that the choice of the loop type depends on the problem to be solved, even if the three available techniques are virtually identical. The **for** loop is useful in cases where you have to repeat something a number of times. This number can be different each time the loop is executed, but looping will stop when the number of repetitions is reached. The **while** and **do-while** loops are executed until the condition that closes the loop becomes false. The difference between the **while** and **do-while** loops is the moment of testing the loop exit condition. For a **while** loop this is done at the beginning of the loop, whereas the exit condition of the **do-while** loop is tested at the end. This implies that this loop is executed at least once, unlike the **while** and the **for** loop.

6.6.1 for

The **for** loop is used like this:

```
for (initialisation; condition; iteration)
{
  …
}
```

Here `initialization` represents the statement or compound statement that is executed when the program enters the loop; `condition` corresponds to the exit condition or rather the condition that must remain true to stay in the loop and `iteration` corresponds to the statement or compound statement to be executed at the "end" of each loop. All three fields may contain any valid C expression and may even be left empty (but make sure that the two semicolons are always there). This explanation shows that the **for** loop is not as elementary as it tries to make you believe. Its most common appearance, and that's how I will use it in this book, is like this:

```
for (int ocd=0; ocd<5; ocd++)
{
  check_that_the_stove_is_off_before_leaving_the_house();
}
```

In this loop, the variable ocd plays the role of loop iteration counter. It can be declared before the loop or, as I did here – even though programming purists do not like this – in the loop's declaration (or header). Note that in this case, the variable ocd does not exist outside the loop. Often the loop iteration counter starts at zero, like here, but this is not at all obligatory. The counter can be incremented or decremented with a constant or a variable or a function. Before each loop iteration the counter is compared to a value that must not be exceeded. If the condition is true, for example because the loop iteration counter is less than the end value, the statements between {}, the loop's body, are executed. If the condition is false, program execution resumes right after the loop. The loop iteration counter is updated after the statements in the loop's body have been executed. ocd++ means that the counter's new value will be its current value increased by one. This is another way to write `ocd = ocd + 1` or `ocd += 1`. Here you could also have written ++ocd, which amounts to the same in this context (I said *this* context!). In C, there are often several ways to achieve the same result, but be aware of subtleties that may not always be obvious.

6.6.2 while

The **while** loop is very similar to the **for** loop despite its very different syntax:

```
while (condition)
{
  …
}
```

As you can see, there are no initialization or iteration fields here, only a condition field. This does not mean that the other two fields are unnecessary (although they can be), it only means that there is no predefined moment when they are executed or evaluated. It is up to the programmer to decide when and where he wants to initialize the loop and maybe increment a loop iteration counter. The condition field can again contain any valid C statement or compound statement but, contrary to the **for** loop, it may not be left empty. A constant can be enough; a non-zero value is interpreted as true, while a zero is considered false. Often a comparison is used, but the result returned by a function call is also usable. The loop's body will only be executed if the condition is true. The three points indicate the statements to repeat.

The following example shows a **while** loop that is identical to the example given above for the **for** loop, except for one detail: the variable ocd exists outside the **while** loop:

```
int ocd = 0
while (ocd<5)
{
  check_that_the_stove_is_off_before_leaving_the_house();
  ocd++;
}
```

The **while** loop is often used to wait for an event, like this:

```
while (he_did_not_say_he_was_sorry==TRUE)
{
  ignore_his_messages(ALL);
}
```

By the way, did you notice the inverted logic of this example? This kind of practice is very risky because it can easily mislead the reader of your code.

When you just need to wait for an event without doing anything, the loop's body, i.e. the part between and including the braces, may be omitted. If you do that, you must replace it by a semicolon ";", like this:

```
while (he_said_please==FALSE) ;
```

6.6.3 do-while

The difference between the **do-while** loop and the **while** loop is the moment of evaluation of the condition that determines whether or not to stay in the loop. For a **while** loop this is done at the beginning of the loop, whereas the **do-while** condition field is evaluated at the end of the loop. A second, more subtle difference is the semicolon that is needed to complete the declaration of the **do-while** loop. For the rest, the two are identical, and everything that I have said before on `condition` is also valid here.

```
do
{
  …
}
while (condition);
```

The **do-while** loop is most useful when it is necessary to perform one or more statements at least once, but perhaps a few times more if a certain condition is true. For example:

```
do
{
  look_both_ways_before_crossing_the_street();
}
while (I_want_to_stay_alive==TRUE);
```

That said, a **do-while** loop can always be transformed into a **while** loop if you make sure that the condition is true at the start:

```
I_want_to_stay_alive = TRUE;
while (I_want_to_stay_alive==TRUE)
{
  look_both_ways_before_crossing_the_street();
}
```

The inverse is of course also true and can sometimes be beneficial to the readability of your program. Keep this in mind!

6.7 More Keys

Now that you know how loops are created in C, let's try to put them into practice. For this, I turned the four-key matrix keypad into a 16-key matrix keypad. I added two rows and two columns to make a 4 × 4 matrix.

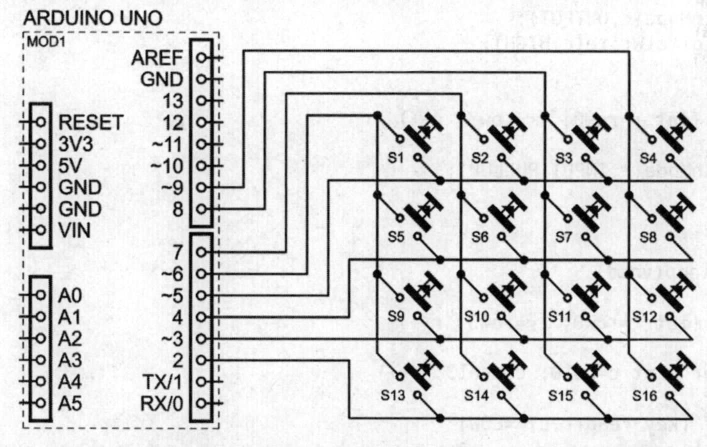

Figure 6-5 - The 16-key matrix keyboard. The buttons are named from top-left to bottom-right.

The sketch that can scan the sixteen keys uses a loop and a function:

```
/*
 * key matrix 2
 */

int row0 = 2;
int row1 = 3;
int row2 = 4;
int row3 = 5;
int col0 = 6;
int col1 = 7;
int col2 = 8;
int col3 = 9;

int key_read(int r, int c)
{
  digitalWrite(c,LOW);
  int key = digitalRead(r);
  digitalWrite(c,HIGH);
  return key;
}

void setup(void)
{
  Serial.begin(115200);

  for (int c=col0; c<=col3; c++)
  {
    pinMode(c,OUTPUT);
    digitalWrite(c,HIGH);
  }

  for (int r=row0; r<=row3; r++)
  {
    pinMode(r,INPUT_PULLUP);
  }
}

void loop(void)
{
  for (int r=row0; r<=row3; r++)
  {
    for (int c=col0; c<=col3; c++)
    {
    if (key_read(r,c)==LOW)
      {
        Serial.print('S');
        Serial.println(4*(r-row0)+(c-col0)+1);
      }
```

```
    }
  }

  delay(100);
}
```

Normally you would start reading a program in its main function, which is the function `loop` in Arduino. Here I wanted to make an exception, because the functions `loop` and `setup` both contain loops, but those of the function `setup` are easier to understand. So let's begin our code review in the function `setup`.

First, the serial port is initialized. We use it to send the name of the button that is pressed to the computer. Then we come across the first **`for`** loop. This loop initializes the outputs that control the four columns of the matrix. I took care to select consecutive pins to facilitate the use of a loop. Now a simple counter can select the outputs one by one. The counter `c` starts at `col0` (pin 6) and counts up to `col3` (pin 9). For each iteration (count) the corresponding pin is configured as an output and it is set to a logic high level. A second loop initialises the inputs, the four rows of the matrix, in a similar manner. The counter r starts at `row0` (pin 2) and counts up to `row3` (pin 5) and configures the corresponding pin as an input with its pull-up resistor activated.

In the function `loop` we find the same two **`for`** loops but this time they are nested. For each value of the row counter, the column loop is repeated four times. This loop is therefore executed sixteen times, once for each key of the matrix. In the column loop the keys are read by a call of the function `key_read`. If this function returns low, it means that the key with the coordinates (`r`,`c`) is being held down. In this case we do a little math to transform the key's coordinates into a key name that corresponds to the circuit diagram. The result is sent to the serial port and can be viewed in the Arduino IDE's serial monitor.

The function `key_read` is simpler than its charlieplexing counterpart because there are no pins to reconfigure. All it has to do is to write and read two pins. The function first activates the requested column `c` before reading the requested row `r`. Then the requested column is disabled again to put the matrix back in its initial state. Finally, the function returns the logic level that was read on the input.

6.8 Ghostbusters

The more keys a keyboard has, the more likely it will be that the wrong key is pressed or that several keys are pressed at the same time. The MCU cannot do much against incorrect keystrokes, but the system can properly deal with simultaneous keys presses. However, a system based on the basic matrix keyboard that we discussed before, cannot. Let me explain.

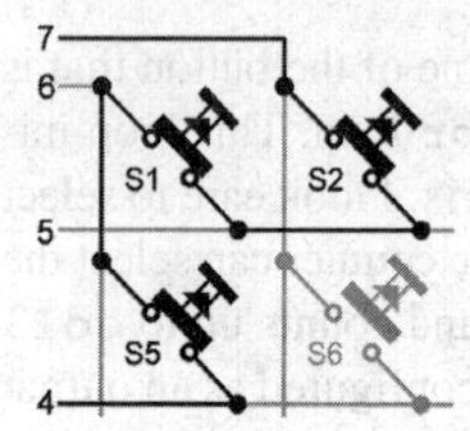

Figure 6-6 - A ghost key. If you simultaneously press the keys S1, S2 and S5, the MCU will believe that not only S2 is pressed, but also S6. The path that induces the MCU in error is drawn in black.

Look at the matrix keypad shown here. This is the same keyboard as before, but I deleted from the drawing the keys unnecessary to this explanation. Remember that pins 6 and 7 are configured as outputs while pins 4 and 5 are configured as inputs. Suppose that the MCU has activated pin 7 by putting a logic low level on it. If I now press S2, pin 5 will see a logic low level and the MCU detects the key press as it should. If I push down S1 at the same time, pin 6 too will see a logic low level. This will not be a problem if the MCU has disabled this output (otherwise there may be a short circuit). However, if, with my Joe Bloggs fingers, I also press key S5, the MCU will detect a logic low level on pin 4, and since the MCU activated pin 7, it will think that S6 is pressed, which is not the case. The MCU has detected a so-called ghost key.

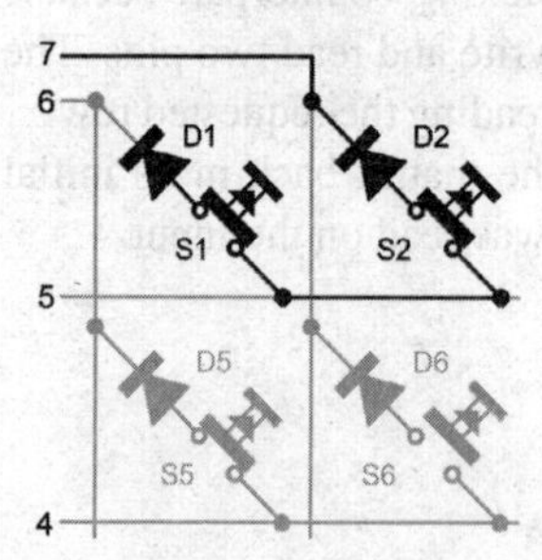

Figure 6-7 - Diodes in series with the keys isolate the rows and columns of the keyboard. There are always two diodes between two columns or two rows that will block the signals.

That's not all. Suppose I now also press key S6 without releasing the other three keys, i.e. I press the four keys at the same time. For the MCU nothing changes and it will keep on detecting key presses on S2 and S6, which happens to be correct in this case. Now I release S2. What will happen? Well, nothing, because the MCU will not notice any differences. The state change of S2 has been made invisible by the other keys; S2 has become a hidden key.

Ghost and hidden keys are a real problem when your fingers are too big for the keyboard or when your fingers fly over the keys very quickly, for example, while playing a video game or the piano or when you write a book. The solution is to connect a diode in series with every key. If the diodes are oriented correctly, they will isolate the columns from each other. The diode will cause a small voltage drop that changes the levels slightly, but this usually does not cause any problems for the MCU.

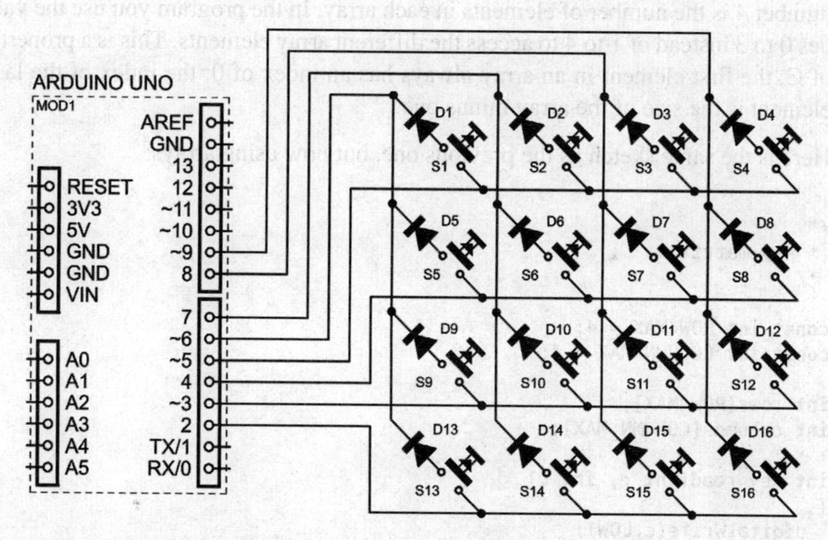

Figure 6-8 - This matrix keyboard is immune to ghost keys.

6.9 Tables

The sketch above that read a 16-key matrix keyboard works well, but it has the inconvenience that the rows and columns of the keyboard must be connected to consecutive pins. Sometimes this is possible, but not always. When it is not possible, the loops to read the keys no longer work and we must modify the program.

Fortunately there is a solution in the form of the table. If we store the pin numbers to use in a table, we can use the same loop to read the keys, but with a slight modification. Instead of representing the pin number directly, the loop iteration counter is now used to indicate the row in the table that contains the pin number to use. This means that there is a little bit more work to do, but you can use any (compatible) pin and the order does no longer matters.

I will explain tables in detail later in this chapter, because I do not want you to lose the thread here. What you need to know now is that by adding `[x]` to a variable name, the latter is converted into a table of x elements. For example:

```
int columns[4];
int rows[4];
```

These are two tables of four integers each. In C, such tables are called arrays. The number 4 is the number of elements in each array. In the program you use the values 0 to 3 instead of 1 to 4 to access the different array elements. This is a property of C, the first element in an array always has an index of 0; the index of the last element is the size of the array minus one.

Here is the same sketch as the previous one, but now using arrays:

```
/*
 * key matrix 3
 */

const int ROW_MAX = 4;
const int COLUMN_MAX = 4;

int rows[ROW_MAX];
int columns[COLUMN_MAX];

int key_read(int r, int c)
{
  digitalWrite(c,LOW);
  int key = digitalRead(r);
  digitalWrite(c,HIGH);
  return key;
}

void setup(void)
{
  Serial.begin(115200);

  rows[0] = 2;
  rows[1] = 3;
  rows[2] = 4;
```

```
  rows[3] = 5;
  columns[0] = 6;
  columns[1] = 7;
  columns[2] = 8;
  columns[3] = 9;

  for (int c=0; c<COLUMN_MAX; c++)
  {
    pinMode(columns[c],OUTPUT);
    digitalWrite(columns[c],HIGH);
  }

  for (int r=0; r<ROW_MAX; r++)
  {
    pinMode(rows[r],INPUT_PULLUP);
  }
}

void loop(void)
{
  for (int r=0; r<ROW_MAX; r++)
  {
    for (int c=0; c<COLUMN_MAX; c++)
    {
      if (key_read(rows[r],columns[c])==LOW)
      {
        Serial.print('S');
        Serial.println(4*r+c+1);
      }
    }
  }

  delay(100);
}
```

As you can see in the function `loop`, the loop iteration counters `r` and `c` in the function `loop` now start at 0 and count up to a maximum value defined at the top of the sketch. Remark how in the comparison the less-than symbol ('<') has replaced the less-than-or-equal symbols ("<=") to prevent the counter from attaining the maximum value which lies just outside the array. Remember, since the counters start at 0, the highest value allowed is the maximum value minus one.

The arguments of the function `key_read` are now values from the arrays instead of the counter values themselves. The calculation to find the key number is now easier, but only because I took a shortcut made possible by the way the circuit has been wired. A better way would have been to create another table for the names of the keys.

In the function `setup` we first fill the two tables. Then the loop counters `r` and `c` are used to read the pin number data from the tables before configuring them as inputs and outputs.

The function `key_read` has not been changed at all.

At the top of the sketch some constants are defined, recognizable by the reserved word **`const`**. "Constant variables" cannot be changed in the sketch and to distinguish them from normal variables I always write them in uppercase. It is imperative here to use **`const`**, otherwise the compiler will not accept the table declarations based on these constants.

6.10 LED Mini-Display

If in the ghost-busted matrix keyboard we short-circuit the keys and if we replace the diodes by LEDs, we obtain a sixteen-point (pixel) display that we can control in a manner similar to that of the matrix keyboard. To avoid that the MCU must provide too much power, we will turn on only one LED at a time. This is not really a problem when we quickly multiplex the sixteen LEDs as our eyes are too slow to notice the trick. With such a display we can create small fun animations. Remember to put a resistor in series with each row or column to limit the current through the LEDs to, say, 5 mA. This may seem little, but it is enough for my eyes. If you find that this is not enough, you are free to increase the current (as long as you stay within the limits of the MCU and the LEDs).

Here is a chase lights sketch that flashes the LEDs one after the other.

```
/*
 * LED matrix 1
 */

const int ROW_MAX = 4;
const int COLUMN_MAX = 4;

int rows[ROW_MAX] = { 2, 3, 4, 5 };
int columns[COLUMN_MAX] = { 10, 11, 12, 13 };
int leds = 1;

void led_write(int r, int c)
{
  digitalWrite(c,LOW);
  digitalWrite(r,HIGH);
  delay(50);
  digitalWrite(r,LOW);
```

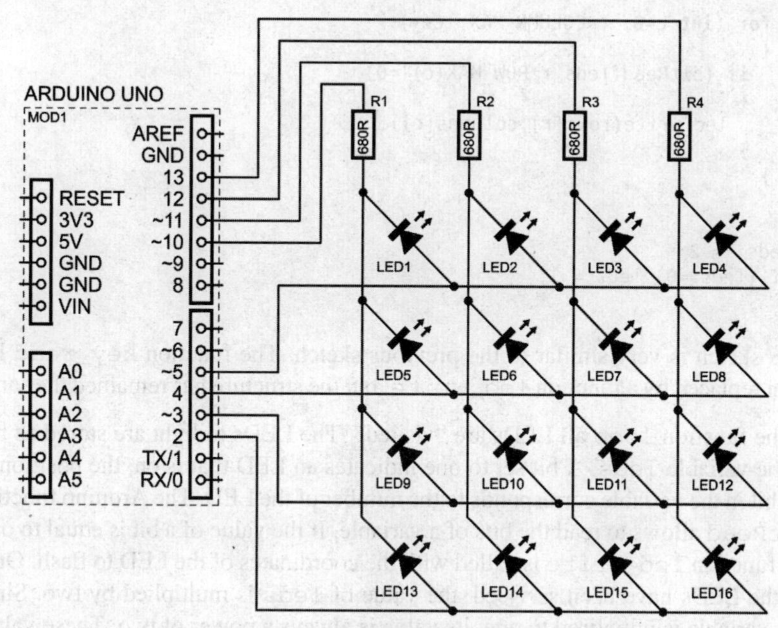

Figure 6-9 - The cousin of the matrix keyboard: the matrix (LED) display.

```
  digitalWrite(c,HIGH);
}

void setup(void)
{
  int i;

  for (i=0; i<ROW_MAX; i++)
  {
    pinMode(rows[i],OUTPUT);
    digitalWrite(rows[i],LOW);
  }

  for (i=0; i<COLUMN_MAX; i++)
  {
    pinMode(columns[i],OUTPUT);
    digitalWrite(columns[i],HIGH);
  }
}

void loop(void)
{
  for (int r=0; r<ROW_MAX; r++)
```

```
  {
    for (int c=0; c<COLUMN_MAX; c++)
    {
      if (bitRead(leds,r*ROW_MAX+c)!=0)
      {
        led_write(rows[r],columns[c]);
      }
    }
  }

  leds *= 2;
  if (leds==0) leds = 1;
}
```

This sketch is very similar to the previous sketch. The function `key_read` has been replaced by a function `led_write`, but the structure has remained the same.

In the function `loop` all LEDs are “visited”. The LEDs to light are stored as bits in the variable `leds`. A bit set to one indicates an LED that is on, the position of the bit in the variable corresponds to the number of the LED. The Arduino function `bitRead` allows to read the bits of a variable. If the value of a bit is equal to one, the function `led_write` is called with the coordinates of the LED to flash. Once all the LEDs have been serviced, the value of `leds` is multiplied by two. Since this variable is initialised to one, its value is always a power of two. These values have the particularity that only one bit is set to one (see table below), all the others are 0, and therefore only a single LED will be lit. The largest power-of-two that fits in a 16-bit integer is $2^{15} = 32,768$. When you multiply this value by two the result will not fit in the variable `leds`, an overflow will occur, and only the 16 lower bits of the result are kept. Since these bits are all zero, the value of the variable `leds` becomes zero. When this happens it is reset to one so that the animation can restart.

The inputs and outputs of the MCU are configured in the function `setup`. The tables are no longer initialized here, because they are now initialized right after their declaration at the top of the sketch.

The function `led_write` lights the specified LED for the time determined by the argument of the call of the function `delay`. This time also controls the speed of the animation.

Power of 2	Decimal	Binary
2^0	1	00000001
2^1	2	00000010
2^2	4	00000100
2^3	8	00001000
2^4	16	00010000
2^5	32	00100000
2^6	64	01000000
2^7	128	10000000

Table 6-1 - The correspondence between the powers of two and the bits of a byte. Any value can be expressed in binary notation by breaking it into a series of powers of two.
For example, $55 = 32 + 16 + 4 + 2 + 1 = 2^5 + 2^4 + 2^2 + 2^1 + 2^0 = 110111$.

6.11 The Parade

The LED matrix can display all kinds of animations; you are limited only by your imagination (and the number of LEDs). For example, if you fill an array with simple 4 × 4-pixel images, you can create an animation by displaying one image after another. A scrolling marquee, i.e. a scrolling text, is another nice option. Here's how to make a one-character scrolling marquee that scrolls the word "ARDUINO" from right to left on our LED matrix.

We begin with the creation of eight 4 × 4-pixel characters ("ARDUINO" plus a space) that we store in a table. Each character occupies 16 bits. The table below shows how I encoded the characters.

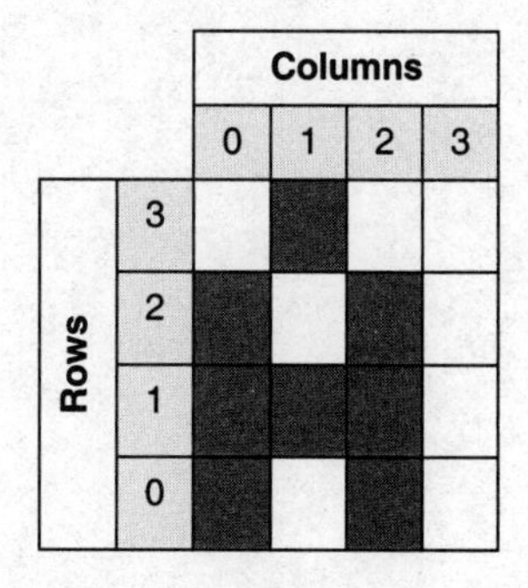

Table 6-2 - 4 × 4-pixel character encoding.
Each cell corresponds to a bit of a 16-bit word. The bit at the crossing Rows = 0, Columns = 0 is at position 0, the one at the intersection R = 3, C = 0 is at the third position and the one at the junction (3,3) is at position 15.
The character 'A' is encoded as follows (bit 0 is on the right): 0b 0000 0111 1010 0111 (In C, binary values begin with "0b").

Each character consists of four columns of four pixels. To scroll through the characters, you can move columns from right to left. When the first column of the first character leaves the left side of the display, the first column of the second character must enter the right side of the display, and so on. Once the second character is shown in full, the sequence restarts, but this time with the second and third characters. To manage scrolling, it is enough to keep track of the index of the first column to display; we can calculate the rest from there.

```
/*
 * LED matrix 2
 */

const int ROW_MAX = 4;
const int COLUMN_MAX = 4;
const int TEXT_MAX = 8;

int rows[ROW_MAX] = { 2, 3, 4, 5 };
int columns[COLUMN_MAX] = { 10, 11, 12, 13 };
int text[TEXT_MAX] =
{
  0b0000011110100111,
  0b0000010110101111,
  0b0000011010011111,
  0b0000111100011110,
  0b0000000011110000,
  0b0000011110001111,
  0b0000011010010110,
  0b0000000000000000
};
int text_index = 0;
int c_offset = 0;
unsigned long int t0;

void led_write(int r, int c)
{
  digitalWrite(c,LOW);
  digitalWrite(r,HIGH);
  delay(1);
  digitalWrite(r,LOW);
  digitalWrite(c,HIGH);
}

void setup(void)
{
  int i;

  for (i=0; i<ROW_MAX; i++)
  {
    pinMode(rows[i],OUTPUT);
```

```
    digitalWrite(rows[i],LOW);
  }

  for (i=0; i<COLUMN_MAX; i++)
  {
    pinMode(columns[i],OUTPUT);
    digitalWrite(columns[i],HIGH);
  }

  t0 = millis();
}

void loop(void)
{
  for (int c=0; c<COLUMN_MAX; c++)
  {
    int ch;
    int c1 = c + c_offset;
    if (c1<COLUMN_MAX)
    {
      ch = text[text_index];
    }
    else
    {
      c1 -= COLUMN_MAX;
      if (text_index+1>=TEXT_MAX)
      {
        ch = text[0];
      }
      else
      {
        ch = text[text_index+1];
      }
    }
    for (int r=0; r<ROW_MAX; r++)
    {
      if (bitRead(ch,c1*COLUMN_MAX+r)!=0)
      {
        led_write(rows[r],columns[c]);
      }
    }
  }

  unsigned long int t = millis();
  if (t-t0>200)
  {
    t0 = t;
    c_offset += 1;
    if (c_offset>=COLUMN_MAX)
    {
      c_offset = 0;
```

```
        text_index += 1;
        if (text_index>=TEXT_MAX)
        {
          text_index = 0;
        }
      }
    }
}
```

Besides managing the display, the function loop also takes care of column scrolling. Since scrolling is based on the columns, I inverted the row and column loops. The outer loop now handles the columns; the inner loop does the rows. Before running the inner loop, a small calculation is done to determine which column should be displayed. The variable c_offset added to the column counter c gives the number of the column. If this number is less than the maximum number of columns (4), the column is still part of the current character. If it is not, then it is a column of the next character. The character containing the column in question is loaded into the variable ch before being displayed. Before loading the character, it is necessary to ensure that the index of the next character is valid or the display may show rubbish.

After refreshing the display, the function loop updates the scroll counter c_offset. This is done every 200 ms, which determines the scroll speed. We cannot use the function delay for controlling the speed, as this would make the animation invisible. Indeed, while the function delay is waiting for the delay to expire, the program is blocked and the display will remain black. Display refreshing must continue while we wait the next moment to scroll. We can achieve this by measuring time using the Arduino function millis. This function returns the number of milliseconds since program start (after a reset). By comparing the current value of millis with the reference value stored at the launch of the sketch, we can wait a well-defined number of milliseconds. When the waiting period is over, we store the current milliseconds value as the new time reference for the next iteration, which allows us to wait again 200 ms.

The function setup is identical to that of the previous sketch, except for the last line that is used to initialize the time reference for the scroll speed regulator. The function led_write has not been changed at all.

After this playful introduction it is now time to get serious.

6.12 A Little Scam

To become an expert on Arduino or, for that matter, on microcontrollers in general, you have to invest in your education. Buying this book – a very good decision, by the way – and an Arduino board probably made a dent in your savings so maybe you would like to earn a little money to get some of your expenses back? Yes? I knew it, and I also know how to do it using the knowledge that you have just acquired.

There is a game where the player tries to find a winning combination by placing four coins on a grid of 4 × 4 positions in such a way that each row and each column contains only one coin. There are 24 possible ways of doing this (4 × 3 × 2 × 1), but only one is the winning combination. If the player succeeds, he/she earns twice the money he/she played, but if he/she gets the combination wrong, the money is lost. After each (winning) game, the combination is changed. You can make the game more difficult by allowing multiple pieces per column (keeping the one-coin-per-row rule) to get 256 possible combinations, but you must increase the prize money accordingly.

An electrified grid is made from non-touching horizontal and vertical metal wires where metal parts placed on a junction connect a horizontal and a vertical wire. The grid can detect all by itself if the winning combination was formed. Does this ring a bell? Exactly, it is a matrix keyboard!

We can make the game more attractive by adding an LED at each junction, which will not only provide visual feedback to show that a coin has been detected at the position where it was placed, but also to display the winning combination to show the player who just lost his money that there really was one. Pressing a button generates a new winning combination. If you are a good DIYer, you can mount the game in a nice box that will accept coins of different sizes (but preferably two euro coins, slightly larger than the British one-pound coin).

This game is particularly interesting for the operator (you) if he/she manages to run the game without a winning combination and without getting caught. This is also why we have provided the option to display the winning combination. It may be fake, but it will inspire confidence. With such a game you will notice that your friends and family quickly start avoiding you, allowing you to spend all your free time again on programming microcontrollers. So how do we go about?

The easiest way is to use a 4 × 8 matrix, half of which will be used by the keys, or rather the coins, and the other half for the LEDs. Therefore we need 12 digital I/O pins, which is no problem for an Arduino board that has at least 14 of those (or 20

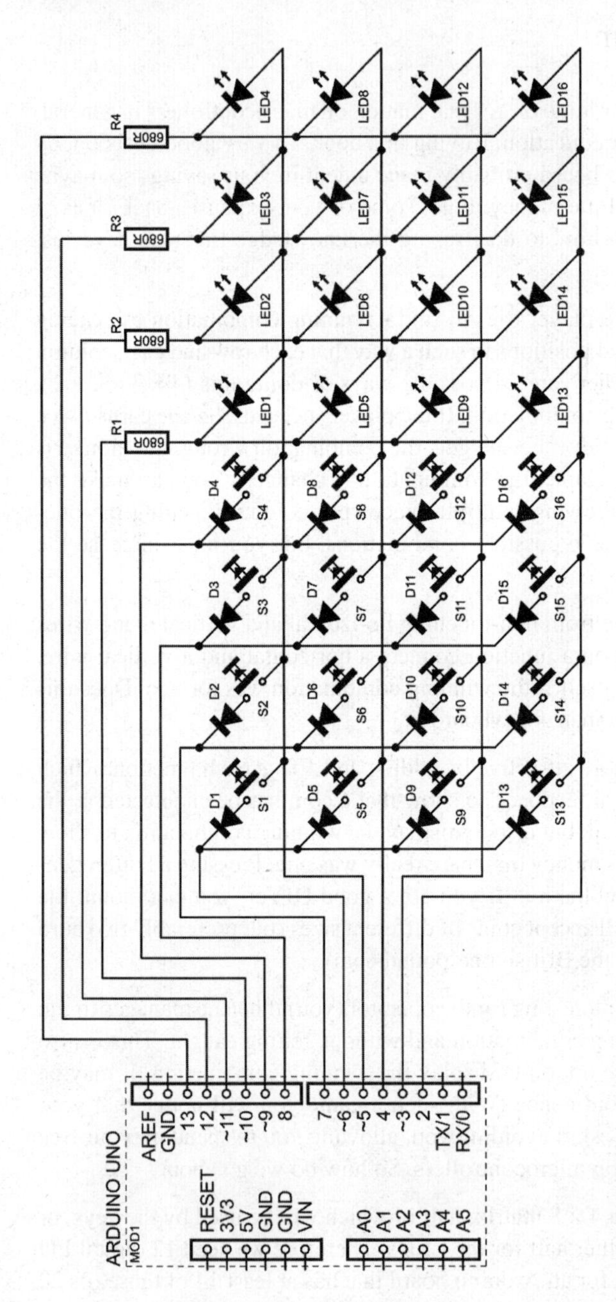

Figure 6-10 - With a matrix keyboard and an LED matrix you can make simple games.

if we also count the analog inputs). It is even possible to keep the serial port's RX and TX signals free, which is convenient because it allows you to connect the game to a PC and record your gains directly into a spreadsheet.

As a pushbutton to restart the game we can simply use the board's reset button. Actually, such a button is not really necessary because the game can reset itself automatically, but it helps to win the player's confidence. A button that does nothing is quite easy to install, but if you choose this option I suggest that you connect it to the board anyway so that it will be credible.

Coins (keys) need to be connected to diodes, since we need a matrix able to correctly detect multiple keys (coins) pressed simultaneously. As explained above, without the diodes this would be impossible. For the same reason, charlieplexing cannot be used here. The LED matrix is identical to the one used for the scrolling LED marquee.

Looking at the circuit diagram you may wonder if it wouldn't have been easier to replace the diodes with LEDs to automatically light the ones that match the coins that have been placed. Indeed, this seems like an attractive option, but it has two drawbacks. First of all, if we do this, the LEDs will be powered by the internal pull-up resistors of the MCU. However, by looking at the datasheet of the MCU, we discover that these resistors have a value between 20 and 50 kΩ, and they therefore cannot provide enough current to fully light the LEDs. External pull-up resistors are needed to remedy this problem. But that's not all. We try to detect a logic low level to determine if a key is pressed or not. An LED creates a voltage drop of approximately 1.6 V and the minimum voltage at an input will therefore be around 1.6 V. However, again according to the MCU's datasheet, for a voltage to be interpreted as a logic low level, it should not exceed $0.3 \times V_{CC}$ which equals 1.5 V assuming V_{CC} is 5 V. If we invert the operating levels of the matrix by inverting the LEDs – a logic high level now corresponds to a pressed key – the situation seems better, since the datasheet specifies that a logic high level is detected starting at $0.6 \times V_{CC}$, meaning 3 V when V_{CC} equals 5 V. I said "seems" because the datasheet also says that an output pin can provide as little as 4.2 V for a logic high level. Subtracting the LED's voltage drop we get $4.2 - 1.6 = 2.6$ V, which is below the detection threshold of a logic high level. In short, if we use LEDs instead of diodes, the matrix will function in an unpredictable manner and even though it may work in some cases, it is not recommended to do so. In order to ensure reproducibility, a good designer always makes sure that the operating range of a circuit does not flirt with its limits and that its functionality never relies on tolerances. A good designer always performs such a worst-case analysis.

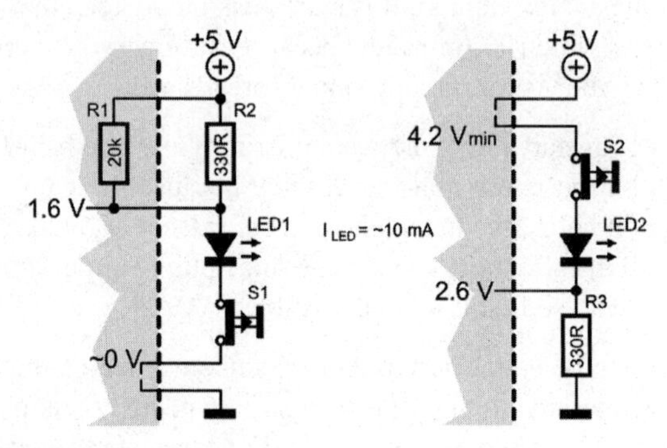

Figure 6-11 - Beware of levels and tolerances.

```
/*
 * money game
 */

const int COLUMN_MAX = 4;
const int ROW_MAX = 4;

int rows[ROW_MAX] = { 2, 3, 4, 5 };
int columns_key[COLUMN_MAX] = { 6, 7, 8, 9 };
int columns_led[COLUMN_MAX] = { 10, 11, 12, 13 };
int keys_pressed[ROW_MAX];
int secret_code[ROW_MAX];
int seed = 0;

void led_write(int r, int c, int d)
{
  digitalWrite(rows[r],HIGH);
  digitalWrite(columns_led[c],LOW);
  delay(d);
  digitalWrite(rows[r],LOW);
  digitalWrite(columns_led[c],HIGH);
}

void led_animation(int repeat)
{
  rows_initialize(OUTPUT);
  for (int n=0; n<repeat; n++)
  {
    for (int c=0; c<COLUMN_MAX; c++)
    {
      for (int r=0; r<ROW_MAX; r++)
      {
        led_write(r,c,20);
      }
```

```
  }
  }
}

void rows_initialize(int direction)
{
  for (int i=0; i<ROW_MAX; i++)
  {
    pinMode(rows[i],direction);
    if (direction==OUTPUT)
    {
      digitalWrite(rows[i],LOW);
    }
  }
}

int key_read(int r, int c)
{
  digitalWrite(columns_key[c],LOW);
  int result = digitalRead(rows[r]);
  digitalWrite(columns_key[c],HIGH);
  return result;
}

void key_read_all(void)
{
  rows_initialize(INPUT_PULLUP);
  for (int r=0; r<ROW_MAX; r++)
  {
    keys_pressed[r] = -1;
    for (int c=0; c<COLUMN_MAX; c++)
    {
      if (key_read(r,c)==LOW)
      {
        keys_pressed[r] = c;
      }
    }
  }
}

void key_show_all(void)
{
  rows_initialize(OUTPUT);
  for (int i=0; i<ROW_MAX; i++)
  {
    if (keys_pressed[i]>-1) led_write(i,keys_pressed[i],1);
  }
}

boolean code_complete(void)
{
```

```
  for (int i=0; i<ROW_MAX; i++)
  {
      if (keys_pressed[i]==-1) return false;
  }
  return true;
}

boolean code_good(void)
{
  for (int i=0; i<ROW_MAX; i++)
  {
      if (secret_code[i]!=keys_pressed[i]) return false;
  }
  return true;
}

void code_show(int period)
{
  rows_initialize(OUTPUT);
  unsigned long int t = millis();
  do
  {
    for (int i=0; i<ROW_MAX; i++)
    {
      led_write(i,secret_code[i],1);
    }
  }
  while (millis()-t<period);
}

void code_generate(int s)
{
  randomSeed(s);
  for (int i=0; i<ROW_MAX; i++)
  {
    secret_code[i] = random(COLUMN_MAX);
  }
}

void setup(void)
{
  rows_initialize(INPUT_PULLUP);

  for (int i=0; i<COLUMN_MAX; i++)
  {
    pinMode(columns_key[i],OUTPUT);
    digitalWrite(columns_key[i],HIGH);
    pinMode(columns_led[i],OUTPUT);
    digitalWrite(columns_led[i],HIGH);
  }
```

```
  led_animation(1);
}

void loop(void)
{
  seed += 1;

  key_read_all();
  key_show_all();

  if (code_complete()==true)
  {
    code_generate(seed);
    if (code_good()==false) led_animation(1);
    else led_animation(7);
    code_show(5000);
    led_animation(1);
  }
}
```

The money game sketch looks quite long because there are many small functions, but some of them have already been described in the previous example. The sketch is not too difficult to understand if you start reading in the function `loop`. At each pass through this function the keys (coins) of the keyboard are scanned (`key_read_all`) and their states are stored. To be precise, the program actually only stores one state per row. Then the LEDs corresponding to the coins are switched on (`key_show_all`). When the four coins are in place, a test performed by the function `code_complete`, the winning combination is generated. It's a bit like playing the lottery, first you fill in your form and then the winning combination is drawn. This is done by the function `code_generate` after which the function `code_good` checks if the winning combination corresponds to the positions of the four coins.

There is a reason for organizing this game as a draw instead of making it a guessing game. For this game we need random numbers, but a microcontroller is a deterministic machine. By design, if you know its initial state, its state at any time in the future can be accurately predicted (if you know all the input signals), so it is the opposite of something random. There are mathematical algorithms capable of producing semi-random numbers, very long number sequences repeated with very long periods, but they suffer from a paradox similar to the chicken and egg causality dilemma: we must start the algorithm in a random state in order to create a new one. If we always start from the same state, the next state will be known in advance. In this sketch we bypass this thorny problem by measuring the time used by the player to place the four coins. This time will never or at least rarely be the same for two consecutive attempts. The sketch measures this delay by counting the number of times the program passes through the function loop until the four coins are in place. The variable seed is the counter and because of the high execution speed of the sketch, its value will increase very quickly and its end value will be pretty random in our case. Although seed occupies only 16 bits and overflows quickly – it will continue at zero – it makes no difference to the random nature of the result. This variable is then used to initialize the random number engine (through calling the function randomSeed in the function code_generate); the random numbers are in turn obtained by calling four times the function random that delivers after every call a random value in the range 0 to COLUMN_MAX - 1, i.e. between 0 and 3.

The draw of the winning combination is announced by a short luminous animation. If the coins are placed at the right positions, a small luminous sequence is shown (led_animation) before displaying the winning combination. If the coins were not on the right spots the sketch only shows the winning combination. This combination remains visible for five seconds and then the game is restarted.

The inputs and outputs of the MCU are configured in the function setup. Since the rows of the matrix must switch between inputs (keyboard) and outputs (LEDs), I created a special function to initialize them on the fly (rows_initialize). A small LED animation indicates that the game is ready.

The function code_generate has been explained above. The function code_show is used to display the winning combination or secret code. To wait a certain amount of time without blocking the display, the same technique is used as in the scrolling LED marquee sketch.

The function code_good compares the secret code to the played code. It is very simple; it just compares two codes character by character. When two different characters are found, the function returns false. The function code_complete is almost identical, but has the opposite behavior: it compares the played code to the code (-1, -1, -1, -1). It returns false when two identical characters are found, meaning that the played code still contains an unidentified key, indicated by a value of -1. The function key_show_all is part of the same family. Like code_complete it compares the played code to the code (-1, -1, -1, -1), but instead of returning false when a key has a value of -1, it lights the corresponding LED if the key has a value greater than -1.

The keyboard is scanned by the function key_read_all. The technique used is the same as the one used at the beginning of this chapter. The only difference is that the numbers of the columns of the pressed keys are stored in the table keys_pressed. This way only one key/coin per column is accepted. As already mentioned, an empty column is indicated by the value -1.

The function key_read is the same as the beginning of the chapter. The function rows_initialize lets you configure all pins at once as inputs or as outputs.

Mentioned several times in the description of the function loop, the function led_animation plays a small animated sequence. It is not very sophisticated, the columns are lit one after the other, I left it as an exercise for you to do better. The function's argument repeat specifies the number of times to play the animation.

We end our review with the function led_write. It is almost identical to the one of the scrolling marquee. Almost, because I added an argument to control the duration of the function. This trick allows the function led_animation to adjust the animation's speed.

6.13 Making New Friends

In the money game sketch, as in all other code examples in this book, I have tried to write all the reserved words or keywords of the C language using boldface characters. Looking carefully at all the functions you will find two that are of type boolean (code_complete and code_good), a word that is not printed in bold. This is normal, because boolean is not a reserved word in C. It is in fact a data type defined in or by Arduino, as are the data types byte and word. The IDE however recognizes these types and shows them in orange, as if they were C keywords. Moreover, the IDE does the same for the API functions it knows of, giving you the impression that the Arduino API is a programming language, which it is not.

I would like to close this example with a remark on my way to name functions and variables. My technique has its roots mainly in laziness. In programming, we often need to edit or copy and rename functions and variables by group. This is easiest when the names differ only by their suffixes or endings, not by their prefixes or beginnings or their middle parts. This technique is even more effective if the text editor allows you to work in columns (a real programmer's text editor can do that). An additional advantage of using suffixes is that it is not necessary to read a name in full when looking for a function or variable, because the beginning is already showing what it is. The result is a bit like the Polish prefix notation in mathematics[1].

1. Not to be confused with the horrible so-called Hungarian notation where the function and variable names are prefixed with mnemonics representing their data type. This can lead to aberrations such as arpszWhatTheHeckIsThis for an array (ar) containing pointers (p) to zero-terminated strings (sz).

6.14 Null Does Not Equal Zero

In Section 6.9 (page 89), I promised you to tell you more about tables. Promise kept, here are some additional explanations. The table or array is very useful because it allows you to treat a data collection without needing to know the names of the individual elements. Text strings for example are actually arrays of characters. A table is created by adding `[x]` to a variable name, where x indicates the number of items that can be stored in the table. To create a multidimensional array just add `[x]` to the variable's name as many times as there are dimensions. In fact, such an array is actually an array of arrays (of arrays, etc.) which are more complicated to use and we leave them aside in this book. Here are some examples of simple arrays:

```
float mona_lisa_dimensions[2];
int mona_lisa_measurements[3] = { 90, 60, 90 };
char mona_lisa_name[] = "Lisa Gherardini";
byte mona_lisa_phone_number[] = { 0, 1, 4, 0, 2, 0, 5, 0, 5, 0 };
```

The first example shows an array of floating point values. The second one shows how to initialize an array of a known size. The arrays in the last two examples do not have a specified dimension since there is no value between the square brackets. Here the size is determined by the number of arguments that initialize the array. So, you wonder, why two examples to show the same thing? Because the third array does not have the size you think it has. Even if "Lisa Gherardini" occupies only fifteen positions, the array can contain sixteen elements. The sixteenth position is reserved for the value 0 (null, not the character '0' or zero) that is added by the compiler to indicate the end of a text string. This is a trick to work around a particularity of C: the number of elements an array can hold is not stored anywhere. Since the null character cannot appear in a string of printable characters, you can use it to signal the end of such a string. This particularity shows both the interest to define constants to determine the size of an array and to USE THESE CONSTANTS.

The first element of an array has index 0 (zero, not null), which is why in C we (almost) always start counting from zero instead of one. The index of the last element of an array is equal to the array size minus one. If you declare an array of ten elements, the last element will have an index of nine. This is why you often see in C comparisons of the "less-than" type instead of the "less-than-or-equal-to" type. Have a good look at the **`for`** loops in the money game sketch above and you will notice that their counters all start at zero and count up to the maximum value minus one.

6.15 Snow White's Apple

As explained in Section 4.5 (page 54), the functions provided by Arduino form an Application Programming Interface or API. APIs are like Lady Apples that have to be eaten greedily without peeling, because the skin contributes heavily to its taste. Did you know that the French call this apple *pomme d'API*? Amazing, isn't it? Even more so once you know that this apple is the oldest known variety, cultivated by the Romans. Maybe the Romans invented programming?

The way you eat a Lady Apple is how you have to use an API. They are designed to be used a maximum without peeling away the skin because that will break code portability, i.e. the possibility of transplanting a program to another platform without much effort. The Arduino API is not an exception, and I advise you to consume it greedily. However, an API not only has benefits and it must be used with care. Actually it is a bit like a poisoned apple.

For example, the functions `digitalRead` and `digitalWrite` are very convenient, but their simplicity has a price: the speed of execution. Indeed, the hardware abstraction provided by the API that allows you to control a pin in Arduino without worrying about the MCU's registers that do the real work, requires several layers of software. This thick layer of code slows down the program and the CPU advances like a spoon being dragged through molasses.

A second disadvantage of the functions `digitalRead` and `digitalWrite` is that they can control only one I/O pin at a time. It is not their fault, they were designed that way, it is due to the Arduino API that does not provide functions to control several I/O pins simultaneously. This omission not only has a negative impact on the execution speed of the program since you must control I/O pins one after the other, but it also affects the synchronization between the I/O pins. With these functions it is simply impossible to activate two outputs or read three inputs at exactly the same time; also using them it is impossible to accurately determine which of two pulses arrived first at the MCU's inputs.

To appreciate the thickness of the API's skin, you have to dive into the source code of Arduino. This is easy since it is included in the Arduino distribution that you have installed. On Windows, the source code of the function `digitalWrite` is found in the file `wiring_digital.c` in the folder <arduino>\hardware\arduino\cores\arduino\ (where you must replace <arduino> by the name of the folder where the `arduino.exe` file lives). Even if you do not understand the code, it is clear that a lot of statements are used to control one pin of the MCU. First the code verifies that a correspondence between the Arduino pin and the pair (I/O port, I/O port pin) exists. It must also ensure that if the pin was busy doing something else, for example outputting a PWM signal, this other function is terminated properly before continuing. Once all this is okay, the state of the I/O pin may be modified, but not without first making sure that this operation cannot be interrupted, otherwise the result will be unpredictable. In total, on my computer, the compiled `digitalWrite` function occupies 102 bytes (`digitalRead` takes 96 bytes), not counting the size of the functions it calls. Note that these are bytes and not machine instructions; also note that probably not all the instructions are executed every time, but this example illustrates how the processor fights its way through a soup of API bytes.

Simplifying this function without losing too much of the security it provides, I managed to get down to 18 bytes, which is almost six times less and it is possible to do even better. This highlights a third problem of the API: it makes your executables fat. This may not necessarily be an issue when the MCU has a 512-kibi program memory, but a memory of 16 kibi is quickly filled if you do not pay attention. This illustrates nicely the importance of working with functions because they appear only once in the executable, but you can call them as often as you like.

6.16 The Core

Now that we have started to peel away the skin of the API, let's go all the way, down to the core. What is the minimum amount of code needed to control a digital output or read an input?

The microcontroller is optimized to handle its I/O pins with the fewest instructions possible. For this it has a register in which each bit corresponds to one I/O pin. A second register determines the data direction of each I/O pin, inbound or outbound. Often a set of special instructions offer a direct access to register bits, which allows you to set, clear or read a bit using a single statement. These instructions need a bit mask to specify which bits must be set or cleared. When such special instructions are not available, you should fall back to the bitwise AND and OR instructions to clear or set one or more bits. Here's how you do this in C:

```
DDRB = 0x30;                    // PB4 and PB5 outputs
PORTB |= 0x20;                  // PB5 = 1
PORTB &= 0xef;                  // PB4 = 0
int pb3 = PINB & 0x04;          // pb3 = PB3
```

Decimal	Hexadecimal	Binary	Decimal	Hexadecimal	Binary
0	0	0000	8	8	1000
1	1	0001	9	9	1001
2	2	0010	10	A	1010
3	3	0011	11	B	1011
4	4	0100	12	C	1100
5	5	0101	13	D	1101
6	6	0110	14	E	1110
7	7	0111	15	F	1111

Tableau 6-3 - The correspondence between decimal, hexadecimal and binary values.

In this example the pins 4 and 5 of I/O port B are configured as outputs, the others are configured as inputs. This is done in the first line. DDR means Data Direction Register and the B indicates that it belongs to port B. The value 0x30 is a hexadecimal value (as indicated by the prefix "0x") which corresponds to bits 4 and 5 (0x30 is 0011 0000 in binary, we count bits from right to left and start at zero). A bit set to one indicates an output, a bit set to zero means an input.

The second line sets bit 5 by executing a bitwise OR operation ('|') with the current value of the register PORTB and the hexadecimal value 0x20 (0010 0000 in binary). The result of this operation is written to the PORTB register. Bit 5 of port B is PB5, i.e. Arduino pin 13, which is connected to the LED on the Arduino board. This line therefore turns on the board's LED.

The third line sets bit 4 to zero by AND-ing ('&') the current value of the PORTB register with the hexadecimal value 0xef (1110 1111 in binary); the result is again written to PORTB. Bit 4 of port B is PB4, i.e. Arduino pin 12, which is not connected to anything. (If you don't believe me, connect an LED to this pin to see for yourself that it stays off.)

The fourth line uses the PIN register of port B to read the logic level on pin PB3 (Arduino pin 11). This is a particularity of the AVR, the state of the I/O pins is read from another register (PINx) than the one that controls these states (PORTx). Reading the PORT register gives you the last value written to this register which not necessarily corresponds to the real pin value. For example, if for some reason an output set to one is shorted to ground, reading its corresponding PORT register will give you one, whereas reading its PIN register will give you the real value of zero.

In the first three lines of the example the PORTB register is first read, then modified, and finally written, in that order. The literature calls this a Read-Modify-Write operation. Three instructions to change the value of a bit do not seem very efficient, especially for an MCU that is supposed to be a specialist in bit juggling, but this is not due to the MCU. The reason for this construction is the C language that does not have functions to set or clear individual bits of a port[1]. Fortunately, the compiler is aware of this weakness and it replaces such constructions by the special instructions provided by the MCU. Thanks to this clever trick, setting or clearing a bit only takes one machine language instruction. Modern compilers are very good in optimizing code; don't waste your time on trying to do better unless you really have to.

To satisfy your curiosity, here is what the lines two and three of the previous example look like in assembly language after passing through the compiler (note that this fragment will not compile in the Arduino IDE):

```
sbi PORTB,5;       // Set BIt number 5 of port B, i.e. PB5 = 1
cbi PORTB,4;       // Clear BIt number 4 of port B, i.e. PB4 = 0
```

That fits in only four bytes. Much better, don't you think so?

6.17 A Trick

`sbi` and `cbi` are nice instructions to play with the pin levels in an efficient way. `sbi` sets the pin level to a logic high, `cbi` makes it a logic low. To toggle the level of a pin, you must first read it and then, depending to the value obtained, choose `cbi` or `sbi`. This is more work than necessary, which is why the MCU provides a shortcut to toggle the level of a pin with a single instruction. The trick is to use `sbi` with the PIN register (the one that is officially read-only) instead of with the PORT register. Every time you do this, the pin level will change value. It's a neat trick because it allows you to generate fast signals or measure the duration of a loop without adding much ballast to the program (only two bytes and two machine cycles). Here's how to use it in a sketch:

1. In C, it is possible to manipulate the bits of a variable in "normal" memory, but not those of a register of a peripheral which is treated as special memory. Modifying a bit of a variable still translates in most cases into a read-modify-write operation, because the processor can only address its memory as (multiples of) bytes, not as bits.

```
/*
 * toggle
 */

int PB5 = 13;                          // PB5 is Pin 13 (the LED)

void setup(void)
{
  pinMode(PB5,OUTPUT);
}

void loop(void)
{
  asm("sbi %0,%1" : : "I" (_SFR_IO_ADDR(PINB)), "I" (5));
}
```

This gibberish[1] produces a square wave signal with a frequency of approximately 570 kHz on pin 13 (PB5, connected to the on-board LED). This sketch also shows that it is very well possible in Arduino to use C functions like **`asm`** that are not part of the Arduino API.

If you want to use another pin (for instance PD2), you must change the value of PB5 (and preferably also its name, to PD2 to continue the example) to make it correspond to the new pin. If the pin belongs to another I/O port you must also change the PIN register (to PIND in this example). Finally, you must change the number of the bit indicated here by `(5)` (to `(2)` to complete the example). Maybe I could have presented the sketch a little nicer, but in that case I must also explain many more things that do not matter here. So, in the end, using an API does have some advantages, doesn't it?

1. This is inline assembly, denoted by **`asm`**`("...")`. The goal is to inject the instruction `sbi PINB, 5` in the assembly code produced by the compiler. The complexity is necessary to avoid a compilation error.

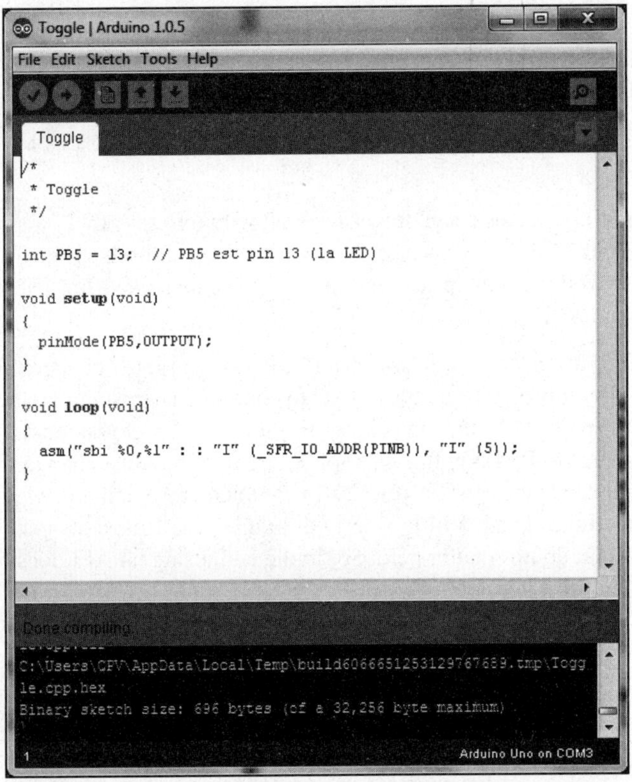

Toggle | Arduino 1.0.5

File Edit Sketch Tools Help

Toggle

```
/*
 * Toggle
 */

int PB5 = 13;  // PB5 est pin 13 (la LED)

void setup(void)
{
  pinMode(PB5,OUTPUT);
}

void loop(void)
{
  asm("sbi %0,%1" : : "I" (_SFR_IO_ADDR(PINB)), "I" (5));
}
```

Done compiling.

```
C:\Users\CFV\AppData\Local\Temp\build6066651253129767689.tmp\Togg
le.cpp.hex
Binary sketch size: 696 bytes (of a 32,256 byte maximum)
```

1 Arduino Uno on COM3

7. Analog Signals: Neither Black Nor White

The previous chapter talked about signals that could take on only two levels – binary signals – and that are handled by digital inputs and outputs. However, in the real world not everything is black or white, there are also intermediate shades of grey and colors. Moreover, very few things are really binary. Dead or alive is one, even if alive can go from almost dead to overexcited as a small child suffering from ADHD. In the world of microcontrollers, signals that in addition to "high" and "low" can take on other levels are considered analog. A three-level signal is as analog as a signal that has an infinite number of levels[1]. To interact with this kind of signals, microcontrollers are equipped with analog-to-digital converters. Digital-to-analog converters produce analog signals. Unlike their counterparts they are not found in many MCUs.

7.1 The Digital Switchover

The Analog-to-Digital Converter or ADC is a device that converts a voltage at its input to a digital value proportional to this voltage. For example, a voltage of 0 V is converted to a value of 0; a voltage of 1 V is converted to 100 and a voltage of 2 V to 200. Since the ADC converts analog values into digital values, the process is also known as digitizing. The ADC is a kind of ruler to measure the "height", i.e. the level, of a voltage and, like a ruler, its accuracy is limited by its scale. If, for example, the scale is in millivolts, you can accurately measure a voltage of 10 mV or 11 mV, but accurately measuring a voltage of, say, 10.3 mV is not possible. The numerical values produced by the ADC are discreet, the ADC quantises the input signal and the analog to digital conversion process is therefore also known as quantization.

The resolution of the analog-to-digital conversion or the quantisation step size, i.e. the smallest difference in the input voltage the AD converter can detect, depends on the number of bits of the converter. 10 bits allow to divide the range of input voltage values into 1024 intervals ($1024 = 2^{10}$, refer to Section 6.10 (page 92) for an explanation of binary numbers), 12 bits offer a resolution of 4096 steps ($2^{12} = 4096$). The MCU of the Arduino board is equipped with a 10-bit ADC, its

---

1. I'll spare you the existential discussion on binary signals that are not really binary. In this book there are binary or digital signals and the others, which are analog.

minimum output value is 0 and the maximum value is 1023 (and not 1024; 1024 in binary representation is 100 0000 0000, which is 11 bits, not 10. Count them if you don't believe me).

According to my dictionary, to measure is to evaluate something in terms of a reference measure. Measuring a voltage with an ADC therefore requires a reference voltage, V_{ref}. For the AVR microcontroller this reference voltage defaults to the supply voltage, which is 5 V for the Uno and Mega boards. This implies that the maximum value of 1023 corresponds to 5 V. For voltages exceeding the reference voltage the ADC will output a value of 1023 (if the input voltage is within the maximum rating of the MCU). To be exact, the maximum value of 1023 corresponds to the reference voltage minus one bit, i.e. 4.995 V, because the range 0 to V_{ref} is divided into 1024 steps (bits) of approximately 4.883 mV and 1023 × 4.883 mV is equal to 4.995 V. In this book we can and we will ignore this small difference. We will study the reference voltage in more detail later on.

As is the case for probably all microcontrollers, the ADC of the AVR has only one input, even if the datasheet seems to suggest otherwise. Although several MCU pins can be used as analog inputs, the ADC can only use one of them at a time. The analog inputs are multiplexed and called channels. The MCU of the Uno has six of these analog channels; the one on the Mega even has sixteen.

The ADC of the AVR is not particularly fast, a conversion takes a minimum time period of 13 µs and reading sixteen multiplexed channels one after the other takes at least 208 µs, which corresponds to a sampling frequency slightly under 5 kHz. According to the Arduino API documentation, a conversion takes about 100 µs, which is seven times slower than the AVR's best conversion time. You will understand that Arduino is not the best platform for audio or video signal processing.

The Arduino API provides two functions to use the ADC in a sketch: `analogRead` and `analogReference`. The first function reads an analog input, you only have to specify the name of the input. The inputs are labeled A0 to A7 for the Uno and A0 to A15 for the Mega, but it is also possible to use the channel numbers 0 to 7 (0 to 15) or the Arduino pin numbers. In this case, you must use 14 to 21 for the Uno and 54 to 69 on the Mega.

If you have read the previous two paragraphs carefully, something must have struck you. Indeed, the MCU of the Uno only has six analog channels whereas the API allows the use of eight channels (A0 to A7). This difference is due to the microcontroller's package. The AVR does contain an ADC with eight channels (it has even more), but its 28-pin DIP package does not have enough pins to allow the connection of all channels. However, the same MCU is also available in 32-pin SMD packages that do expose the two supplementary analog channels. (The other two extra pins of this package are reserved for the power supply.)

Okay, enough talked for now, it's time to start playing with the analog inputs. Here is a voltmeter sketch that reads once per second the analog input A0 and transmits the raw result on the serial port. The result, converted to volts, is also transmitted.

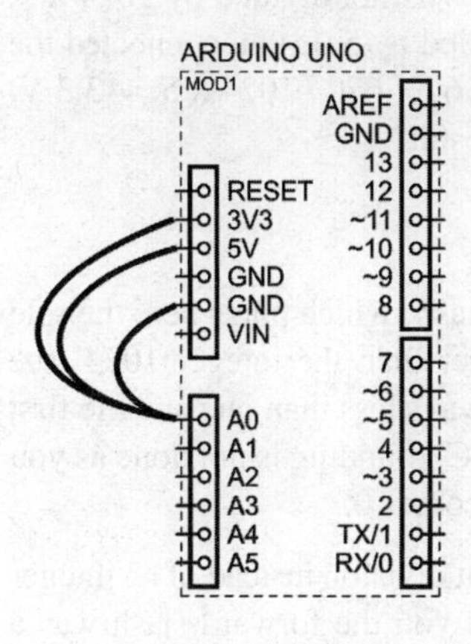

Figure 7-1 - A short wire connects the voltmeter input A0 to one of three voltage sources available on the Arduino board. You can also leave the left end of the wire dangling in the air like a small antenna to pick up noise.

```
/*
 * voltmeter
 */

void setup(void)
{
  Serial.begin(9600);
}

void loop(void)
{
  int v = analogRead(A0);
  Serial.print("A0 = ");
  Serial.print(v);
  Serial.print(" = ");
  Serial.print(v/1024*5);
  Serial.println(" V");
  delay(1000);
}
```

It is not necessary to connect a voltage source to try this sketch, a bit of wire is enough, because the Uno has three voltage sources on-board: 0 V, 3.3 V and 5 V. Additionally, you can use your fingers or other body parts to generate random values. Load the sketch in the board, open the serial monitor of the IDE to view the measured voltage, and then connect the input A0 to one of the mentioned voltage sources, but not to 0 V.

Do you notice anything strange? Good. As you will have seen, the converted voltage in volts is always zero, how can this be possible? The conversion seems to be correct, since the C language respects the same operator precedence in the calculations as the one you learned in school. The value v is not divided by 1024 × 5, but by 1024 and the result of the division is multiplied by 5. If you connected the wire to 3.3 V, you should get a raw value of around 670; 670 / 1024 × 5 ≈ 3,3 V. So then why doesn't the microcontroller find this result?

7.1.1 Type Conversion

The answer lies in the vicious symbol '/' (forward slash) which plays here the role of integer division. The integer `v` is always smaller than the integer 1024 (see above), meaning that the result of the division is always less than one and the first integer smaller than one is zero. 0 × 5 = 0, *QED*. In C, rounding is not done as you would probably do it, and 0.999… as an integer becomes 0.

The problem is easy to solve, just use a floating-point division instead of an integer division. And the function to do this is... '/'! I told you the forward slash was a vicious symbol. In fact, '/' is the division operator but, in C, if the numerator and denominator are both integers, the division will be an integer division (but without remainder). If on the other hand one of the two or both are floating-point values, the division is treated as a floating-point division, even if the quotient is stored in an integer.

There are several ways to ensure that the division type will be the right one, but they amount to the same: use at least one floating-point value to do a floating-point division and use only integers if you need an integer division.

The C language provides the so-called typecast to allow you to temporarily change the data type of a variable. Thus, for example, it is possible to transform for the duration of an operation an integer value in a floating-point value or vice versa.

To temporarily convert a data type to another, just put the new type in front of the variable to convert, like this:

```
float(integer)
(float) integer
int(character)
```

This works for all types, also for those defined by you. Constants can also be forced into types, like this:

- integer: 1024
- floating point: 1024.0

Now that you know how to change the data type of variables, you can correct the sketch above to calculate the voltage properly. Several solutions exist, for example adding "`.0`" to the value 1024. It is also possible to convert v to **`float`** or even declare it as a **`float`**, but this last option also changes the way the sketch prints the raw value.

```
Serial.print(v/1024.0*5);
Serial.print(float(v)/1024*5);
Serial.print((float)v/1024*5);
Serial.print((float)v/float(1024)*5);
```

7.1.2 The Bulk of the Budget is Spent on Representation Costs

Now that the voltage is calculated correctly, you may want to try the sketch with a potentiometer. But wait a moment; wouldn't it be more interesting to turn the sketch in a multichannel voltmeter first? This is easy because the channel numbers are consecutive and so a simple counter is enough to read them all.

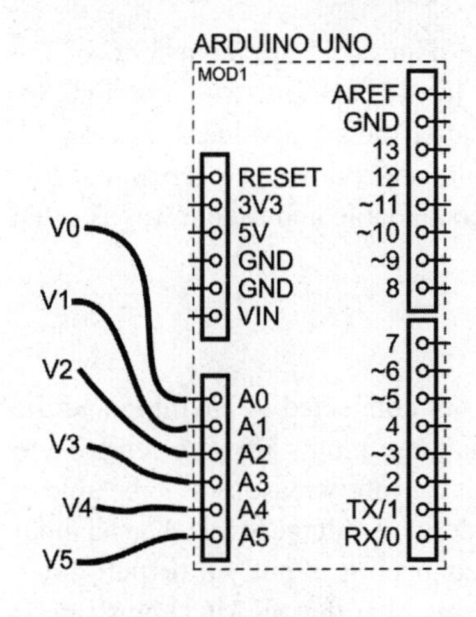

Figure 7-2 - A simple six-channel voltmeter.

```
/*
 * multichannel voltmeter
 */
```

```
void setup(void)
{
  Serial.begin(9600);
}

void loop(void)
{
  for (int i=0; i<6; i++)
  {
    Serial.print("A");
    Serial.print(i);
    Serial.print("=");
    Serial.print(analogRead(i)/1024.0*5);
    Serial.print("V");
    if (i<5) Serial.print(", ");
  }
  Serial.println(".");
  delay(1000);
}
```

This sketch generates once per second frames like these:

```
...
A0=2.80V, A1=2.50V, A2=2.38V, A3=2.38V, A4=2.38V, A5=3.27V.
A0=2.84V, A1=2.37V, A2=2.03V, A3=1.79V, A4=1.67V, A5=3.27V.
A0=3.10V, A1=2.81V, A2=2.54V, A3=2.29V, A4=2.01V, A5=3.27V.
...
```

Have you noticed how the presentation of the data takes up about 80% of the sketch? To make the data look pretty, I wanted to end each line with a period. So, in order to prevent the loop to print a comma after the sixth value, I used an `if` statement to ensure that only the first five results are followed by a comma. It is often like that; presenting data in a clean, understandable and usable way is often more difficult that producing it.

7.1.3 A Tip

In the previous sketch the analog inputs were not connected to anything and the readings fluctuated randomly. This is normal as these inputs are very sensitive to parasitic signals present in the air. It may be that for safety reasons or to be able to detect anomalies your application needs a well-defined voltage level when an input is not connected. This is easy to achieve by connecting a pull-up or pull-down resistor to such an input. But there is a trick. Remember that all MCU pins can be digital inputs and they all have a built-in optional pull-up resistor. When you configure a pin as an analog input it can be used as a digital input at the same time, thus allowing you to activate the pull-up resistor. If you do this the voltage read on an open input will be the supply voltage, 5 V in the case of the Uno and Mega boards.

7.1.4 ADC References

As I said above, the Arduino API provides two functions for analog inputs. Now that you know how to use the first one, `analogRead`, let's move on to the second: `analogReference`. This function allows you to select the ADC's reference voltage. Why? Well, in many cases the voltage to convert never reaches the maximum level (the power supply) and the upper range of the possible numeric values remains unused. This is a waste and it is not good for the signal-to-noise ratio of the conversion. The solution is to adapt the reference voltage to the input signal.

The function `analogReference` requires one argument that can have the following values:

- `DEFAULT`: the reference is the supply voltage;
- `INTERNAL`: the reference is 1.1 V (for the Uno);
- `INTERNAL1V1`: the reference is 1.1 V (Mega only);
- `INTERNAL2V56`: the reference is 2.56 V (Mega only);
- `EXTERNAL`: the reference is the voltage (between 0 V and the supply voltage) applied to the pin AREF.

Be very careful with the external reference. If the AREF pin is connected to an external voltage source, the internal references should not be used otherwise you risk damaging the MCU. Its datasheet is very clear: if an internal reference is selected, it will be shorted to the external reference. This apparent weakness of the MCU actually allows you to use the internal reference outside the device, through the AREF pin. If you would like to try this, make sure to buffer the pin, because the output can only provide very little current. The Arduino documentation suggests putting a resistor of a few kΩ in series with the external reference to allow on-the-fly reference switching. I'm not sure if that is such a good idea, because the internal reference may be influenced by the external reference. I do not recommend using this technique if your application needs to perform highly accurate measurements.

7.2 Back to Analog

To be able to measure analog signals is one thing, but often it is also desired to generate analog signals. However, like many MCUs, the AVR does not have a Digital-to-Analog Converter (DAC) to do this. So how can it be it that the Arduino API features a function `analogWrite`? Because Arduino uses a trick.

This trick is called Pulse Width Modulation (PWM) and is well known as it is more or less imposed by the MCU manufacturers. Why? Well, to find out I contacted some microcontroller manufacturers and it turned out that, according to them, the DAC eats up too much space ("real estate") on the chip. That space is expensive and chip builders prefer to use it for (multi-channel) PWM devices. Such devices are very similar to counters and timers, modules that the manufacturer puts in the MCU anyway, and so he often creates a hybrid timer/counter peripheral that can do PWM too.

A PWM signal is rather easily converted into an analog signal by using a low-pass filter (see also Section 3.2.7, page 28). In many cases it is not even necessary to filter the PWM signal, because the device to control often behaves itself as a low-pass filter. A motor for instance can be driven very well by a PWM signal, because its mechanical inertia acts as a low-pass filter. And if the device to control does not suffer from mechanical inertia, it is you who will play the role of the low-pass filter! Take for example an LED. Controlling the brightness of an LED with a PWM signal is very common. Unlike a motor, an LED does not suffer from mechanical inertia, it is actually extremely fast, but your brain is too slow to detect the modulation and you will see a constant brightness level instead. You are a low-pass filter (with a pretty low cut-off frequency as well).

Another reason to sometimes prefer PWM signals to analog voltages is to save energy. A transistor controlled by a PWM signal is either blocking (no current passes through it) or conducting (creating a low voltage drop and therefore dissipating only a little bit of power). It spends very little time in intermediate states where the real energy losses occur. Spending as much time as possible in these two extreme states limits the energy consumed by the transistor.

Arduino's `analogWrite` function produces a PWM signal instead of a voltage; it is up to you to turn it into a real analog signal if necessary. The duty cycle of the signal is determined by its 8-bit argument, 0 corresponds to 0% (a constant low level) and 255 to 100% (a constant high level). Because the PWM signal is produced by the MCU's counters, `analogWrite` only works with those pins that are internally connected to the counters. The number of PWM output pins available depends on the MCU, the one on the Uno has six of them (pins 3, 5, 6, 9, 10 and 11, marked on the board by a tilde '~'), the Mega has fifteen (pins 2 to 13, an entire row, and also pins 44 to 46 that are not marked).

Here is a sketch that controls the brightness of an LED connected to pin 11 (this is not the on-board LED) by a voltage on the analog input A0.

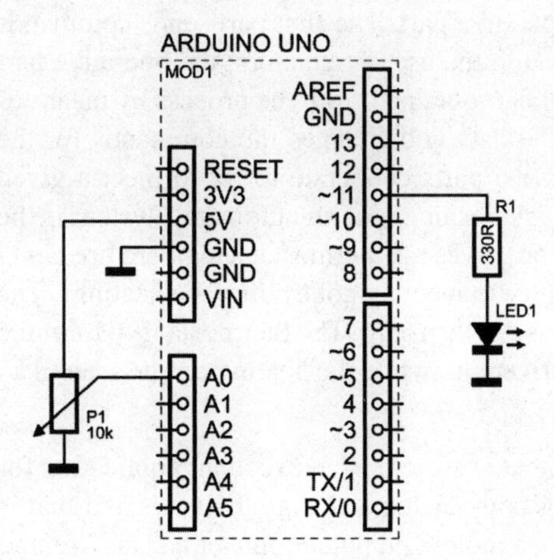

Figure 7-3 - The potentiometer controls the brightness of the LED.

```
/*
 * dimmer
 */

int led = 11;

void setup(void)
{
}

void loop(void)
{
  analogWrite(led,analogRead(A0)/4);
}
```

This sketch shows that it is possible in C to combine multiple statements in a single line. The value provided by `analogRead` is divided by four to ensure that it does not exceed 255, the maximum value `analogWrite` can handle.

7.3 Look Ma, No Hands!

As I mentioned before, motors are often controlled with PWM signals. A field of electronics where there are many systems that control one or more motors (pumps, fans, conveyors, robots, etc.) is automation. Control engineers build so-called automated systems capable of controlling a process autonomously. Such a system consists of a command part and an operative part. The first part sends commands to the second who executes them through the use of actuators; the operative part sends status information to the control part obtained from the process by means of sensors. The control unit processes the data and changes the commands for the operative part accordingly. Thus the two parts cooperate to try to meet a given point of operation set by the operator. An example of an automated system is the heating in your home. The operator (you) gives the thermostat a temperature target of, say, 20 °C that the thermostat will try to meet by controlling the heating[1]. The process here is the (heating of the) house or the room. The thermostat is part of the control unit, but also part of the operative unit, while the heating can be seen as an actuator.

As you will have understood, an automated system is an excellent application for a microcontroller capable of handling both analog and digital signals and that is why I decided to show you how to turn a motorized potentiometer and an Arduino board into an automated system. Controlling a motorized potentiometer may seem exotic, but this type of application is actually very common in automation, as it all comes down to controlling an actuator (a motor) as a function of a signal obtained from a sensor (a potentiometer) to attain a target set by the operator (you). The motorized potentiometer is a good simulation of such a system, because its behavior is more complex than you (may) think. The type I used, bought on the internet, is a spare part for a motorized audio mixer, but you can also make one yourself. Linear or rotating, it does not matter; you can also replace it with some other voltage source directly or indirectly affected by a motor.

7.3.1 Motor Driver

Control engineers have developed beautiful mathematical theories to fully specify and determine the behavior of an automated system. The major drawback of these theories is that, to be of any use, control engineers must rigorously model the pro-

1. In this system, if you play the role of the thermostat the system is no longer automated, even if you're willing to go back and forth between the couch and the heater to adjust the temperature. Man has no place in automated systems, except for giving instructions.

Figure 7-4 - The motorized fader with its driver.

cess to be automated. However, and this is where the theory falls short, in general such a model cannot be created. In practice the process dynamics and disturbances are only known (very) roughly, the system often behaves in a complex way and its parameters may vary over time. Finally, the tasks to be performed by the controller are usually poorly specified. In short, as often in this world, theory and practice are two different things. Fortunately, control engineers have developed techniques that help to achieve satisfactory results with poorly modelled systems. One of these techniques is to arm yourself with a Proportional-Integral-Derivative (PID) regulator, the step response of the open-loop system and the Ziegler-Nichols method.

Let's start with the step response of the open-loop system. As you will see, it is not as complicated as it sounds. Simply disconnect all feedback paths in the system (open the loop), switch on the actuator for a while and observe the result. For my motorized potentiometer this means connecting one end of the potentiometer to 0 V and the other end to, say, 5 V, connecting a voltmeter to the wiper (set to its zero position), switching on the motor briefly (while making sure that it spins in the right direction) and recording how the voltage on the wiper changes (careful, it moves fast!). Doing all this manually is not really a solution, we will have to automate this process, and we can do that with our MCU board and a motor driver.

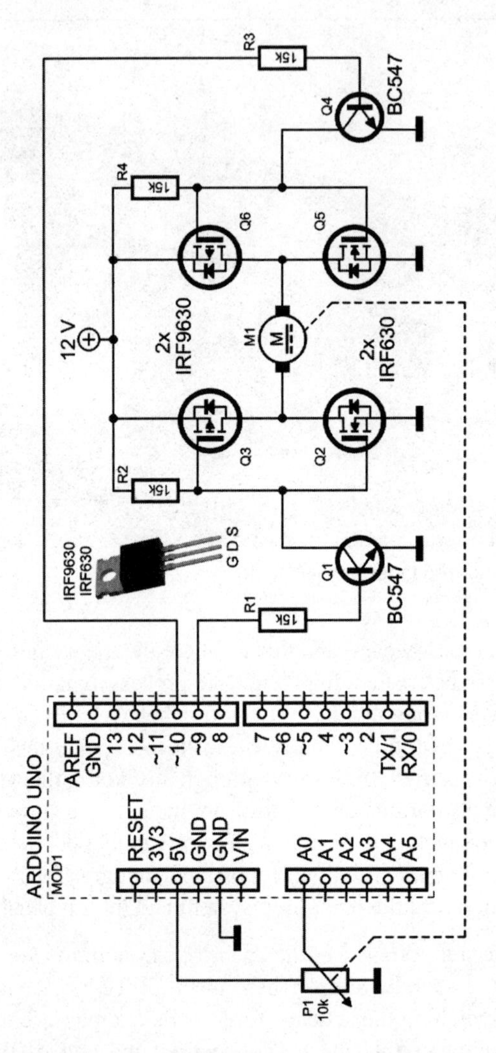

Figure 7-5 - The driver of the motorized potentiometer is a classic H-bridge. With the transistors used the bridge is capable of switching 100 W, more than enough for the little motor that consumes just 2 W.

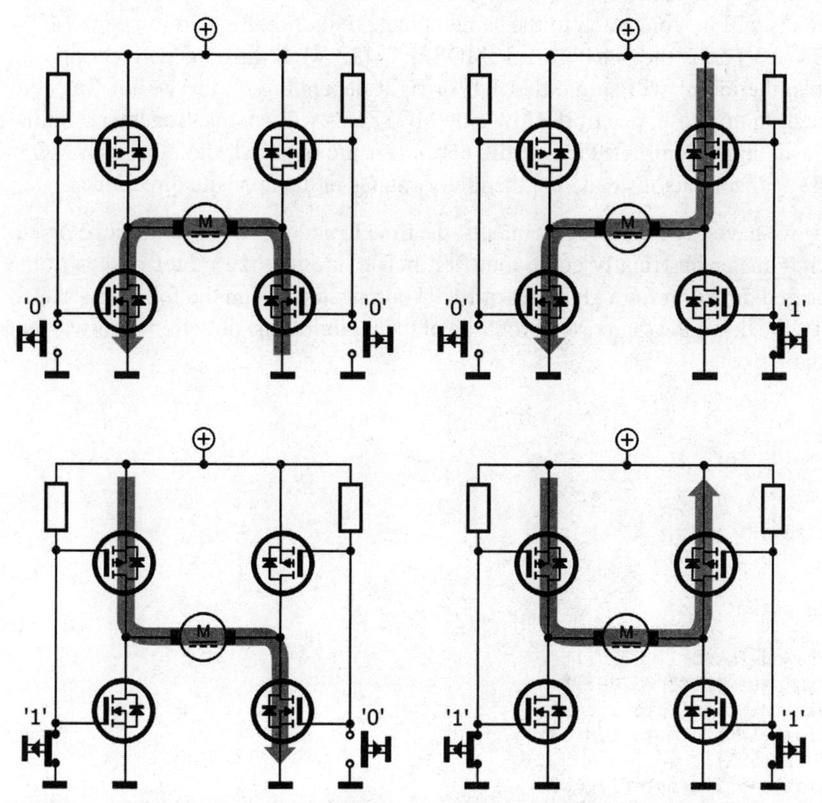

Figure 7-6 - The four modes of an H-bridge. The '0' and '1' states in each drawing indicate the logic levels to be provided by the two outputs of the Arduino board to activate the selected mode.

The potentiometer wiper is connected to an analog input of the MCU, its little motor (see photo above) is connected to two digital outputs of the Arduino board through some transistors that provide the power necessary to run the motor in both directions. This same circuit will be used without changes later on as our automated system with the loop closed within the microcontroller, in software.

For the motor driver I chose a classic H-bridge, built from MOSFETs and powered from 12 V. The bridge is controlled by two MCU ports fortified by transistors (Q1 and Q4). These transistors are needed to avoid contact with the 12 V of the H-bridge; the MCU cannot handle more than 5 V on its pins. Two ports define four states, but two states are identical here: stop, left, right and stop. If both ports output the same logic level (both high or both low) the motor will not rotate, as its contacts will be connected to the same voltage, either by the two low-side MOSFETs (0 V) or by the two high-side MOSFETs (12 V). If the ports deliver opposite levels, the motor will spin either left or right depending on the way it was connected. In this case, one of the low-side MOSFETs will conduct together with the opposite high-side MOSFET. If the port states are inverted, the other two MOSFETs will conduct instead, and the motor rotates in the opposite direction.

Now we have to do some experiments, the time to get out your oscilloscope or frequency meter has finally come, but first make sure that the motor rotates in the expected direction for a given set of MCU output states. Run the following sketch and write down the direction of rotation of the motor (or the direction of movement of the wiper).

```
/*
 * motor test 1
 */

int motor1 = 9;
int motor2 = 10;

void setup(void)
{
  pinMode(motor1,OUTPUT);
  pinMode(motor2,OUTPUT);
  digitalWrite(motor2,LOW);
  digitalWrite(motor1,HIGH);
  delay(500);
  digitalWrite(motor1,LOW);
}

void loop(void)
{
}

My motor rotates: to the left / to the right / not at all
```

Did you notice that the function `loop` is empty? The function `setup` takes care of everything here: it defines the state of the outputs and switches the motor on for 500 ms. If you need more time to see which way the motor spins, increase this value.

If the motor is not running, look for the problem. Did you provide the 12 volts for the H-bridge? Did you wire the bridge correctly? Are the transistors connected as they should be? Maybe you inverted them (the N-MOSFET must be connected to 0 V; the P-MOSFET must be connected to 12 V)? Etcetera.

7.3.2 Obtaining a Step Response

We will now modify and extend our sketch to test the analog input, like this:

```
/*
 * motor test 2
 */

int motor1 = 9;
int motor2 = 10;
int slider = A0;

void setup(void)
{
  Serial.begin(115200);
  pinMode(motor1,OUTPUT);
  pinMode(motor2,OUTPUT);
  pinMode(slider,INPUT);
  digitalWrite(motor1,LOW);
  digitalWrite(motor2,LOW);
}

void loop(void)
{
  Serial.println(analogRead(slider));
  delay(250);
}
```

The motor is at rest and every 250 ms the MCU sends the value measured on analog input A0 over the serial port. Vary the voltage to make sure it is digitized. Write down the minimum and maximum values if the voltage does not go all the way to 0 V or 5 V or to neither of them.

Now it is time to get serious and use the oscilloscope or frequency counter. The frequency to be measured will be around 2 kHz. In the following sketch (`Motor test 3`) comment out the ***<u>bold, italicized and underlined</u>*** line using double forward slashes "//", like this:

```
  // i += 1;
```

Cut the power to the motor driver. Load the sketch in the MCU, connect the oscilloscope or frequency counter to pin 13 and write down the frequency of the observed signal. This is the frequency with which we are going to sample the step response of the open-loop system.

Sample frequency: ____________ Hz

```
/*
 * motor test 3
 */

int PB5 = 13;                    // PB5 is pin 13 (the LED)

#define SAMPLES_MAX 512
int samples[SAMPLES_MAX];
int i = 0;
boolean off = FALSE;

int motor1 = 9;
int motor2 = 10;
int slider = A0;

void setup(void)
{
  Serial.begin(115200);
  pinMode(PB5,OUTPUT);

  pinMode(motor1,OUTPUT);
  digitalWrite(motor1,LOW);
  pinMode(motor2,OUTPUT);
  digitalWrite(motor2,LOW);
  pinMode(slider,INPUT);

  i = 0;

  digitalWrite(motor2,HIGH);
  digitalWrite(motor1,0);
}

void loop(void)
{
  asm("sbi %0,%1" : : "I" (_SFR_IO_ADDR(PINB)), "I" (5));

  int s = analogRead(A0);
  s += analogRead(A0);
  s += analogRead(A0);
  s += analogRead(A0);
  s /= 4;

  if (i<SAMPLES_MAX)
  {
    samples[i] = s;
    i += 1;
  }
```

```
  if (off==FALSE && s>=600)
  {
    digitalWrite(motor1,LOW);
    digitalWrite(motor2,LOW);
    off = TRUE;
  }

  if (i>=SAMPLES_MAX)
  {
    Serial.println(off);
    for (int j=0; j<i; j++)
    {
      Serial.print(samples[j]);
      Serial.print(',');          // or use '\t' instead of ','
    }
    Serial.println();
    Serial.println("done");
    while (1);
  }
}
```

Now uncomment the ***bold, italicized and underlined*** line and, while keeping the motor driver unpowered, program the sketch in the MCU. It will switch the motor on and keep it running until the value of the digitized voltage exceeds 600. If this value is too high, change it before continuing.

Open the Arduino serial monitor. Set the wiper of the potentiometer to its minimum position, press the board's reset button and keep it pressed down. With your other hand, turn on the power to the motor driver. Release the reset button; the motor should run for a short while, and the wiper should move. Wait until the numbers in the terminal stop scrolling[1]. That's it. If all went well, you managed to capture the step response of your system.

It may be that your system is slower than mine and that the sampling rate is too high. If this is the case, there will not be enough memory to store the complete step response and the loop must be slowed down. This can be achieved by adding more `analogRead` lines. If your system is really slow, use the function `delay`. Do not forget to remeasure the sampling rate if you modify the speed of the loop.

Use the copy and paste feature of your operating system to copy the values displayed in the terminal to a spreadsheet and save the file. The spreadsheet is an excellent tool for visualizing data in graphical form. Here is the step response I obtained.

---

1. Be careful; use this method only with low power motors (or systems). I assume no liability for any injury or damage caused by this method.

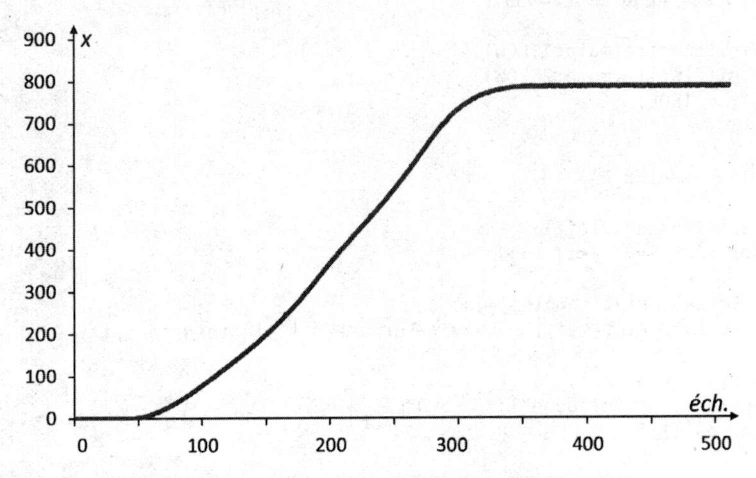

Figure 7-7 - The step response of the system described in Section 7.3.1.

Let's now take a little break to study the sketch in detail. As usual we start reading in the function `loop`. Here we first encounter the gibberish function introduced in the previous chapter to measure the speed of the loop or the sampling frequency. Then analog input A0 is read four times. This is done for two reasons: first, it helps to smooth the raw data and second, it slows down the loop rather dramatically. The second reason seems a little strange, but the technique is surprisingly effective. Why? Because in Arduino, the ADC clock runs at 125 kHz and a conversion takes 13 cycles. All the statements that together make up the function `analogRead` limit the execution frequency of `loop` to about 9 kHz. With four `analogRead` in series I measured a sampling frequency of 2.2 kHz.

After taking a sample, if there is space left, the sample is stored.

Next follows the test to see if it is time to stop the motor. This test consists of two conditions. The first condition, implicating the variable `off`, is to prevent that the test is true more than once. The goal here is to keep the loop execution speed during sampling as constant as possible by avoiding the execution of unnecessary statements. The second condition fixes the target value for our step response at 600. This value was chosen somewhat at random, but not entirely, because it prevents the wiper from bumping into the end of the potentiometer. However, if the target value is too small, the wiper will not reach its maximum speed. This value therefore is a compromise.

When the memory is full, sampling stops because the program will enter a dead-end from where the recorded data is sent over the serial port before the sketch hangs in an infinite loop. How to format the data is up to you, I chose a Comma-Separated Values (CSV) format because that is well-supported by spreadsheets. Depending on the spreadsheet it may be more convenient to separate the samples by a tab "\t" or a semi-colon instead of by a comma. You can do that here (see also the comment in the sketch).

After the function `loop` now have a look at the function `setup`. In this function everything is normal and known; the only thing to remember here is that the motor is switched on at the end of this function. The delay between the end of `setup` and the first pass through `loop` is negligible.

7.3.3 The Compound `if`

Using multiple conditions in an **`if`** statement was a novelty in the previous sketch. You can put as many conditions as you want, but it is not very good for the readability of your program. The conditions are separated by Boolean operators like AND ("&&") and OR ("||"). For a NAND or NOR operation you must invert the condition using the NOT ('!'). Parentheses are used to group the Boolean operations.

```
if (I_will_stop_eating_sweets==TRUE && I_will_eat_less_fat==TRUE
           && !(I_will_stop_jogging==TRUE || I_will_drink_more==TRUE))
{
  I_may_be_hit_by_a_truck_and_die();
}
else
{
  I_may_collapse_under_my_own_weight_and_die();
}
```

All conditions are not necessarily evaluated. In C, as soon as a condition ensures that the result of the complete test cannot be changed by any of the following conditions, the program stops evaluating the remaining conditions. The order of evaluation is always from left to right, respecting of course the parentheses. For example:

```
if (tar_water_does_nothing==TRUE || snake_oil_has_no_effect==TRUE)
{
  Take_an_alka_selzer();
}
else
{
  Get_back_to_work_you_lazy_bum();
}
```

If the first condition is true the second condition will not be evaluated, because as soon as one of the terms of an OR operator becomes true, the result will be true as well[1]. This can be a problem if one or more of the conditions contain function calls (programming purists frown on this) that are not executed because a condition cancelled the evaluation of the conditions next in line. On the other hand, this may also be the desired operation. In short, pay attention to the order of conditions.

7.3.4 The PID Controller

Let us now return to the step response, the graph that tells us so much about the nature of a system (to automate or not). What the experts already knew, we only discover after consulting the literature: the step response I recorded is that of a time-delay system probably consisting of several first-order subsystems (the motor, the wiper) and some nonlinearities (including friction). This may sound scary, but we have two powerful weapons that allow us to master this kind of graphs without deep knowledge: the PID controller and the Ziegler-Nichols method.

We will start with the first, the PID controller. In short, such a device controls a process by monitoring its current state of mind (P), its tendencies (I) and its moods (D). To better understand the PID controller, let's study the example of a car speed regulator or cruise control. Our goal is to drive a car at a constant speed. For this we monitor the car's speedometer and we try to keep the needle at the desired value by pushing down or releasing the throttle. This type of control is based on the current or instantaneous speed.

Another method to keep a constant speed is to ensure that the acceleration remains zero. A car that does not accelerate or decelerate moves at a constant speed (or is not moving at all, which is also a constant speed). A third way to control the speed is to measure the distance traveled per period of time. If this distance is the same for periods of the same duration, the speed of the car must be constant.

The way you press the throttle to keep the instantaneous velocity constant is a proportional (P) response: you watch the speedometer and according to the observed movements of the needle you press the accelerator or you release it. Now remember your physics lessons. The acceleration $a(t)$ is equal to the change in velocity divided by the change in time, i.e. dv/dt. In other words, the acceleration is the derivative (D) of the velocity. For the distance traveled (position) we can write

1. Which is a pity, because now we still do not know if snake oil has an effect or not.

$x(t) = \int v(t)\mathrm{d}t$, meaning that it is the integral (I) of the velocity. If we combine these three techniques into a single control device, we obtain a PID controller. To summarize: the PID controller controls a system using three related criteria.

The 'P' of proportional is nothing more (or less) than an amplifier or attenuator (in the example of the cruise control the gain depends on how well you master your right foot), the 'I' of integral is a low-pass filter and the 'D' of derivative is a high-pass filter. Together they guide the system to keep it on the right track. Consult a book or the internet to learn the exact theory of PID controllers and their more or less gifted brothers and sisters. In the meantime I will show you how to implement a PID controller in only a few lines of code.

```
/*
 * non working PID controller
 */

char motor1 = 9;
char motor2 = 10;
char slider = A0;
const float Kr = 1.95;          // proportional gain.
float yi[2] = { 0.0, 0.0 };     // low-pass filter 1.
float ki[2];
float yd[2] = { 0.0, 0.0 };     // low-pass filter 2.
float kd[2];
int target = 500;               // target value.

float lpf(int sample, float coeffs[], float y[])
{
  y[0] = coeffs[1]*sample + coeffs[0]*y[1];
  y[1] = y[0];                  // delay for 1 sample period.
  return y[0];
}

void setup(void)
{
  pinMode(motor1,OUTPUT);
  digitalWrite(motor1,LOW);
  pinMode(motor2,OUTPUT);
  digitalWrite(motor2,LOW);
  pinMode(slider,INPUT);
}

void loop(void)
{
  int error;
  float spd_p;
  float spd_d;
  float spd_i;
```

```
  int spd;
  int val = analogRead(slider);

  // calculate the error.
  error = target - val;

  // calculate the proportional part (P).
  spd_p = error*Kr;
  // calculate the integral part (I).
  spd_i = lpf(error,ki,yi);
  // calculate the derivative part (D).
  spd_d = error - lpf(error,kd,yd);
  // add P, I & D.
  spd = spd_p + spd_i + spd_d;

  if (spd<0)
  {
    // Clockwise.
    digitalWrite(motor1,HIGH);
    analogWrite(motor2,-spd);
  }
  else
  {
    // Anti-clockwise.
    digitalWrite(motor2,HIGH);
    analogWrite(motor1,spd);
  }
}
```

Warning, this sketch does not work, it only shows the principle. Missing here in particular are the filter coefficients for the integrator (I) and differentiator (D) and a few details to adapt it to the real world.

We start reading the sketch in the function `loop` that begins by sampling the current state. This state is compared to the target value, which gives the error that has to be corrected. By multiplying the error by a constant (obtained from the model of the system that we do not know; we will come back to that later), we find the proportional component. The same error passed through a low-pass filter provides the integral component, and after filtering it with a high-pass filter we get the derivative component. These three components are added together to form the correction signal that has to be applied to the system. It is that simple. Here the correction is a speed which may be positive or negative, the sign of which must be converted to a rotation to the left or to the right.

Note how the functions `digitalWrite` and `analogWrite` are mixed freely without reconfiguring pins. This is possible because these functions configure themselves.

Figure 7-8 - The automated system in all its splendor.

7.3.5 The Digital Filter

As you may have noticed, the previous sketch only contains a low-pass filter, the high-pass filter is obtained by subtracting the output of the low-pass filter from the input signal. The theory of digital filters is not covered in this book. If you want more information, refer to the literature. The theory of digital filters is complex, but their implementation is not too involved. Because digital filters are used in many microcontroller applications, I decided to show you here very briefly how to use them.

The low-pass filter used above is a first order filter (6 dB / octave), coded as follows:

$$y[n] = (1-k)x[n] + ky[n\text{-}1] \text{ where } k = e^{-2\pi f_c f_s}$$

and f_s = sample frequency
f_c = cut-off frequency

This is an Infinite Impulse Response digital filter (or IIR). Because the calculated value $y[n]$ depends in part on the previous result $y[n-1]$ and since k cannot be zero, the filter's response to an impulse will be infinite. With the above equations, a sampling frequency of 2.2 kHz and a cut-off frequency of, say, 21 Hz, we find for k a value of 0.94178726829….

7.3.6 Dynamic Duo

We are now in possession of almost all the pieces to complete the PID controller and its sketch. Still lacking are the coefficients of the two filters and the value for the proportional gain K_r. The Ziegler-Nichols method will provide us with these values by examining the step response of our system. This method is simple and gives good results. Look at the drawing below, it shows us all we need to know. All you have to do is draw a few lines and measure some distances.

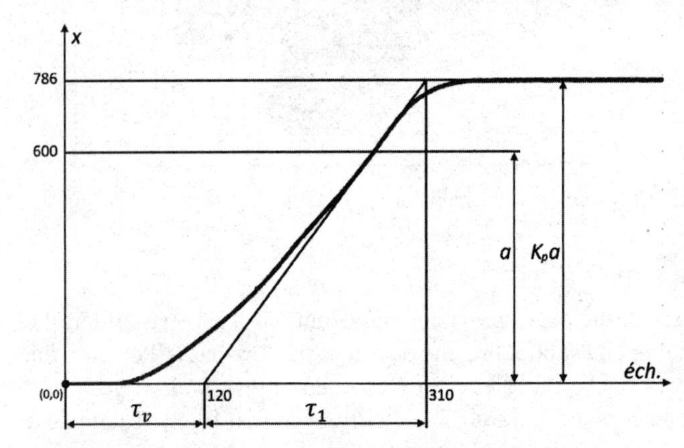

Figure 7-9 - The Ziegler-Nichols method offers an easy way to obtain the values needed to correctly setup a PID controller.

According to the Messrs. Ziegler and Nichols, for a process with a step response like the one shown above and that must be controlled by a PID controller, the time constants and the gain are given by:

- $K_r = 1.2\ K_p \times \tau_1 \tau_v$
- $\tau_i = 2\tau_v$
- $\tau_d = 0.5\tau_v$

The tangent is supposed to pass through the inflection point of the curve. The curve I drew is not a pure step response, because I did not know exactly when to stop the motor while being sure that it had reached its maximum speed. As point of inflection I therefore took the point where the curve crosses the target value of 600, the moment when the motor was switched off. By measuring the distances in the graph

I found a value of 120 samples for τ_v and 190 samples for τ_I. At 2.2 kHz the interval between samples is approximately 455 μs, so we can calculate a value 109 ms for τ_i and 27 ms for τ_d. When we convert these time constants to cut-off frequencies using the equation $f_c = 1 / (2\pi\tau)$, we find that $f_{c,i} \approx$ 1,5 Hz and $f_{c,d} \approx$ 5,8 Hz. Using these results we calculate k_i = 0,995725 and k_d = 0,983572.

Finally, $K_p = 768/600 = 1.31$ and therefore $K_r = (1.2/1.31) \times (190/120) = 1.45$. That's it! We have everything we need and we can finalize the sketch:

```
/*
 * PID controller
 */

char motor1 = 9;
char motor2 = 10;
char slider = A0;

const float Fs = 1760;                  // Hz, sample frequency
const float Kr = 1.45;                  // gain P.
// Filter for I.
const float Fci = 1.5;                  // Hz, cut-off frequency
float ki[2];
float yi[2];
// Filter for D.
const float Fcd = 5.8;                  // Hz, cut-off frequency
float kd[2];
float yd[2];
int target = 500;

void lpf_init(float fs, float fc, float coeffs[], float y[])
{
  coeffs[0] = exp(-TWO_PI*fc/fs);       // k
  coeffs[1] = 1.0 - coeffs[0];          // 1-k
  y[0] = 0.0;
  y[1] = 0.0;
}

float lpf(int sample, float coeffs[], float y[])
{
  // y[n] = (1-k)*x[n] + k*y[n-1]
  // k = e^(-2*pi*fc/fs)
  // fs : sample rate
  // fc : cut-off frequency
  y[0] = coeffs[1]*sample + coeffs[0]*y[1];
  y[1] = y[0];                          // delay for 1 sample period
  return y[0];
}

void setup(void)
```

```
{
  Serial.begin(115200);

  pinMode(motor1,OUTPUT);
  digitalWrite(motor1,LOW);
  pinMode(motor2,OUTPUT);
  digitalWrite(motor2,LOW);

  lpf_init(Fs,Fci,ki,yi);
  lpf_init(Fs,Fcd,kd,yd);
}

void loop(void)
{
  int error;
  float spd_p;
  float spd_d;
  float spd_i;
  int spd;

  // Sample 4 times to slow down the loop to about 2 kHz
  int val = analogRead(slider);
  val += analogRead(slider);
  val += analogRead(slider);
  val += analogRead(slider);
  val >>= 2;

  error = target - val;               // calculate error
  spd_p = error*Kr;                   // P
  spd_i = lpf(error,ki,yi);           // I
  spd_d = error - lpf(error,kd,yd);   // D
  spd = int(spd_p + spd_i + spd_d);   // PID together
  spd = constrain(spd,-255,255);      // apply a speed limit

  // The sign of spd determines the direction.
  if (spd<0)
  {
    // Anti-clockwise.
    digitalWrite(motor2,LOW);
    analogWrite(motor1,-spd);
  }
  else
  {
    // Clockwise.
    digitalWrite(motor1,LOW);
    analogWrite(motor2,spd);
  }

  // Received new target value?
  if (Serial.available()>=3)
  {
```

```
    target = Serial.read() - '0';              // hundreds
    target = Serial.read() - '0' + target*10; // tens
    target = Serial.read() - '0' + target*10; // units
  }
}
```

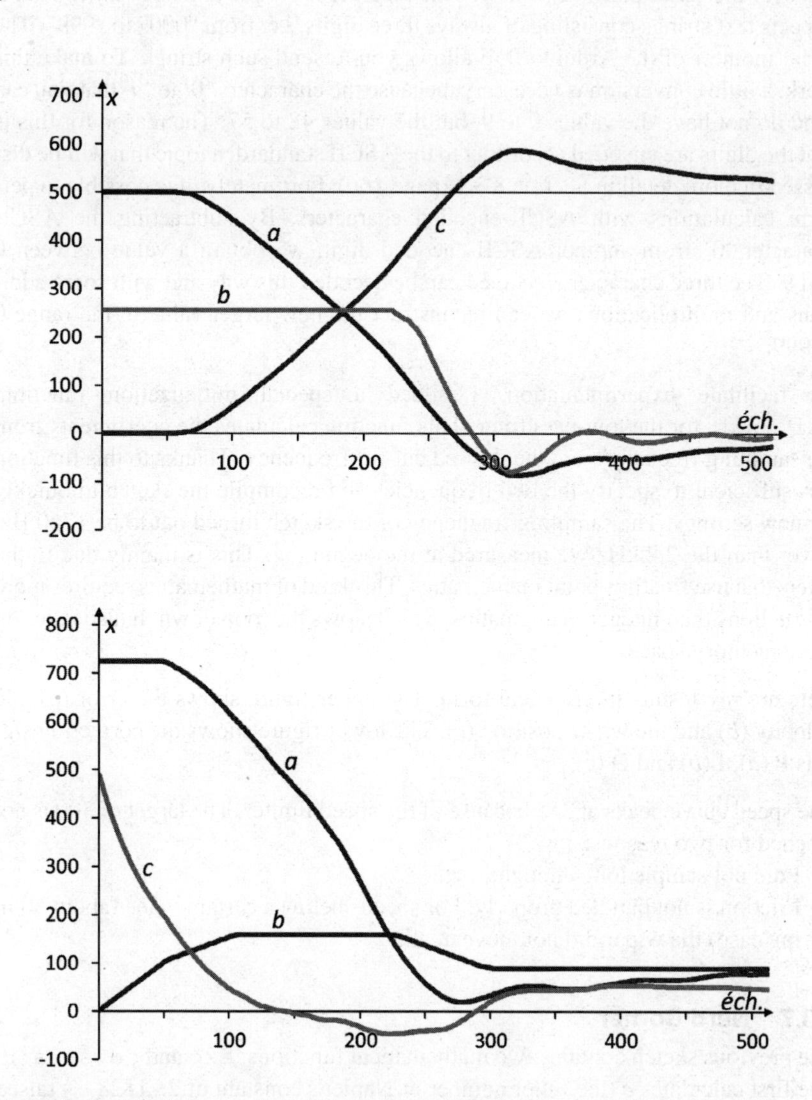

Figure 7-10 - The behavior of the PID system. Left: error (*a*), speed (*b*) and the wiper position (*c*). Right: the correction signals P (*a*), I (*b*) and D (*c*).

After all this work, this sketch does not really contain any surprises. There is a novelty compared to the previous sketch: a speed limiter keeps the velocity in the range of 0 to 255 as accepted by `analogWrite`. The ability to receive target values over the serial port is also new. This receiver is simple and not very robust. It expects text strings consisting of always three digits, i.e. from "000" to "999". The serial monitor of the Arduino IDE allows you to send such strings. To make this work, a little conversion is necessary, because the characters '0' to '9' that you can send do not have the values 0 to 9, but the values 48 to 57. The reason for this is that the digits are encoded according to the ASCII standard, a topic that will be discussed in more detail in Section 8.3.1 (page 166). Fortunately, it is possible to perform calculations with ASCII-encoded characters. By subtracting the ASCII character '0' from another ASCII-encoded digit, we obtain a value between 0 and 9. The three characters received can be decoded this way and with some additions and multiplications we can reconstruct the new target value in the range 0 to 999.

To facilitate experimentation, I added a special initialization function, `lpf_init`, for the low-pass filter. This function calculates the coefficients from the sampling frequency and the desired cut-off frequency. Thanks to this function it is sufficient to specify the two frequencies and recompile the sketch to quickly try new settings. The sampling frequency of the sketch turned out to be 1760 Hz, lower than the 2.2 kHz we measured at the beginning. This is mainly due to the filters that use floating-point mathematics. This kind of mathematics requires more instructions than integer mathematics, which slows the loop down. It also takes up more memory space.

Here are my results in graphical form. The upper figure shows the error (*a*), the velocity (*b*) and the wiper position (*c*). The lower figure shows the correction signals P (*a*), I (*b*) and D (*c*).

The speed curve peaks at 255 because of the speed limiter. The target of 500 is not reached for two reasons:

1. I did not sample long enough;
2. Friction is not handled properly. For speeds below a certain value (about 50 in my case) the wiper did not move at all.

7.3.7 Nerd Corner

The previous sketch contains two mathematical functions: `exp` and `constrain`. The first calculates *e* (the Euler number or Napier's constant of 2.71828...) raised to the power specified between the functions parentheses. The Arduino API provides a similar function called `pow` which takes two arguments instead of one,

Figure 7-11 - A heuristic approach yields near-optimal results.

a base and an exponent, but Arduino does not know the constant e (it does know $\pi/2$, π and 2π). This is why I opted for `exp` instead of defining e myself and use it with `pow`.

The second mathematical function, `constrain`, constrains its argument to be within a range. If the value to constrain is smaller than the lower bound, `constrain` will return the lower bound. If, however, the value to limit is larger than the upper bound, `constrain` will return the upper bound. In all other cases, the value itself is returned.

Unlike `exp`, the function `constrain` is part of the API's group of Math functions. Members of this group are actually not that mathematical as most of them are simple functions that limit values. Taking the square root (`sqrt`) or calculating the power of a number (`pow`) are also part of this group. According to the Arduino documentation, some trigonometric functions are part of the group Trigonometry. By grouping things this way the API makes you believe that these are the only mathematical functions available in Arduino. However, this is not true. Arduino also allows the use of mathematical functions contained in the compiler's math library, as I did above with `exp`. This means that you have access to a function like `hypot` that calculates the hypotenuse, i.e. $\sqrt{}(a^2+b^2)$. Look at the file `math.h` which lives in the folder hardware\tools\avr\avr\include of your Arduino installation. Once again, Arduino is an API, not a programming language.

7.3.8 Sneak Preview

The previous sketch used the serial port to receive a value. This value consisted of three digits, received character by character. With the function `Serial.available` you can find out if characters are available in the serial port's receive buffer, and if so, how many. The function `Serial.read` is used to read the characters one by one from the buffer. It is not possible to read several characters at once. There exists also a function `Serial.peek` that behaves much like `Serial.read` except that it does not remove the character from the buffer. This kind of function is useful when a character must be processed by several functions before it is eaten by a `Serial.read`.

The serial port and methods for manipulating text strings are discussed in detail in the next chapter.

7.4 Recreation: the Misophone

Phew! After this long exercise to implement a PID controller, it is time to relax a bit. For this I propose a nice project that requires a bit of mechanics, electronics and programming.

Have you ever heard of misophonia? Probably not, because it is a little-known disease, identified only in the 1990s. Misophonia, not to be confused with mysophobia, a pathological fear of contamination and germs, is "hypersensitivity to everyday noises, most commonly other people's eating and breathing sounds. This can trigger extreme feelings of rage or panic, or even imagining doing violence to the maker of the sound."[1] When, some time ago, I watched a video on the internet showing an electronic fork that made a sound when the food stuck on it came into contact with the mouth of the user of the fork, I immediately thought of misophonia. The purpose of the musical fork, dubbed EaTheremin by its Japanese inventors, was to encourage children to eat vegetables while having fun. I have my doubts about the effectiveness of this somewhat surprising approach, but such a fork could perhaps help people who suffer from misophonia out of their social isolation. Imagine, equipped with these musical forks, the person suffering from this disease could again invite friends over for dinner at his or her home where the sounds emanating from the mouths of the guests would be rendered inaudible by the harmonious singing of the musical forks. So, how to build such a fork (that I renamed to Misophone)?

1. Source: *A Misophonia UK information leaflet*, ref. M001.3 (www.misophonia-uk.org)

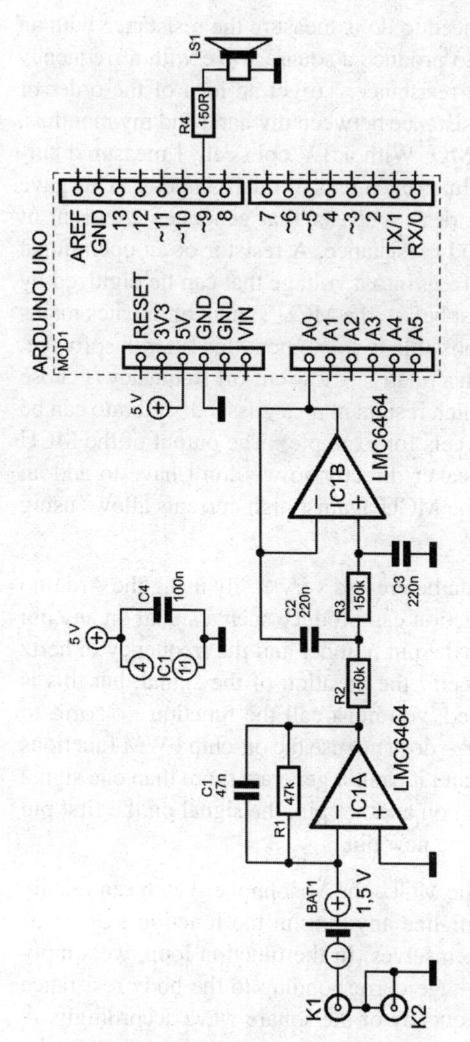

Figure 7-12 - Schematic of the Misophone.

The fork held in the hand of its user, the food stuck on the fork together with the user's body form an electrical circuit that closes when the food touches a conducting spot on the user, preferably the mouth, but a piece of bare skin works as well. The frequency of the sound produced depends on the electrical resistance of the food, but also on the electrical resistance of the body of the user. We therefore need a circuit capable of producing a sound of which the frequency is controlled by a resistor. Such a circuit is pretty easy to make with only "analog" components, but a microcontroller will do too. All we need to do is measure the resistance with an analog input and use a digital output to produce a square wave with a frequency that varies according to the measured resistance. To get an idea of the order of magnitude we are talking about, the resistance between my hand and my mouth is, according to my multimeter, about 1 MΩ. With a 3 V coin cell, I measured currents around ten microamperes, sometimes even tens of microamperes. This gave me the idea of treating the musical fork as a sensor that generates a current of which the intensity is a function of body resistance. A resistor or an operational amplifier (op-amp) can convert the current into a voltage that can be digitized by the MCU. A low-pass filter inserted just before the MCU's input eliminates mains hum picked up by the body. It is also possible to implement this filter in software, simplifying the circuit even more. With a relatively low cut-off frequency (I chose 5 Hz) a certain inertia is introduced which results in nice glissandi. Vibrato can be created with flexible foods (certain sweets for example). The output of the MCU is powerful enough to drive a loudspeaker directly so we don't have to add an audio amplifier. A resistor to protect the MCU against high currents allows using a low-impedance loudspeaker.

The variable frequency square wave can be created very easily using the Arduino function `tone`. This simple to use function can produce such a signal on any pin of the MCU; you only have to specify the pin number and the frequency in hertz of the signal. It is also possible to specify the duration of the signal, but this is optional. If the duration is not specified, you must call the function `noTone` to stop the square wave. The function `tone` does not use the on-chip PWM functions of the MCU, only an on-chip counter, and it cannot generate more than one signal simultaneously. If you call `tone` again on another pin, the signal on the first pin will be removed before continuing on the new pin.

With the filtering carried out outside the MCU, the Misophone sketch can remain very simple. It is not necessary to initialize anything in the function `setup` as `tone` and `analogRead` initialise themselves. In the function loop, we simply call `analogRead` to measure the voltage corresponding to the body resistance before calling `tone` to change the frequency of the square wave accordingly. A test to see if the measured resistance is very high (no contact) allows muting the sound when the food is not in contact with the mouth. `analogRead` provides values from 0 to 1023 that could be used directly to set the frequency from 0 to 1023 Hz but, since the human ear is capable of hearing much higher frequencies, scaling the frequency can improve the audible effect of small variations in body impedance. I chose a scaling factor of 3.

```
/*
 * misophone
 */

int speaker = 9;

void setup(void)
{
}

void loop(void)
{
  unsigned int frequency = analogRead(A5);
  frequency *= 3;
  if (frequency<100) noTone(speaker);
  else tone(speaker,frequency);
}
```

That leaves us with transforming a fork into a controller that can be used as a switch. One of the contacts of the switch is the handle of the fork that is held by the user, the other contact is formed by the teeth of the fork with the food stuck on it. The switch is closed when the user puts the food in his mouth. Here's how I went about to make such a controller.

For the oral contact I used a fork, made from metal of course, with a relatively thin handle. Some left-over copper tubing of about 12 cm (4.75") long and 16 mm (5/8") in diameter (check the plumbing department of your local hardware store), slightly flattened to better match the shape of the handle of the fork, plays the role of hand contact of my fork switch. For the wiring I used flexible audio cable with two shielded conductors recovered from a broken headset, which had a nice stereo plug welded on one end of the cable. Very useful. I first put the cable through the copper pipe, because the plug was too big to do this later on. Using a bit of self-adhesive copper tape stuck on the handle of the fork, I could solder one of the conductors of the cable to the fork. The handle was then insulated with plastic tape. I soldered the shields of the two conductors of the cable together before insulating them and then wrapped the cable around the handle of the fork to prevent it from being ripped off too easily; a piece of tape holds everything in place. Then, after sliding the handle of the fork with the cable into the tube, I soldered the second conductor to the end of the tube. Finally, I filled the tube with hot glue to stabilize the assembly.

And voilà your Misophone is ready. You will now have to invite some friends for dinner during which you will pretend that you suffer from misophonia. This is quite plausible, because this disease can manifest itself at any age. But since you care about your friends, as you tell them, you have prepared a Misophone for each

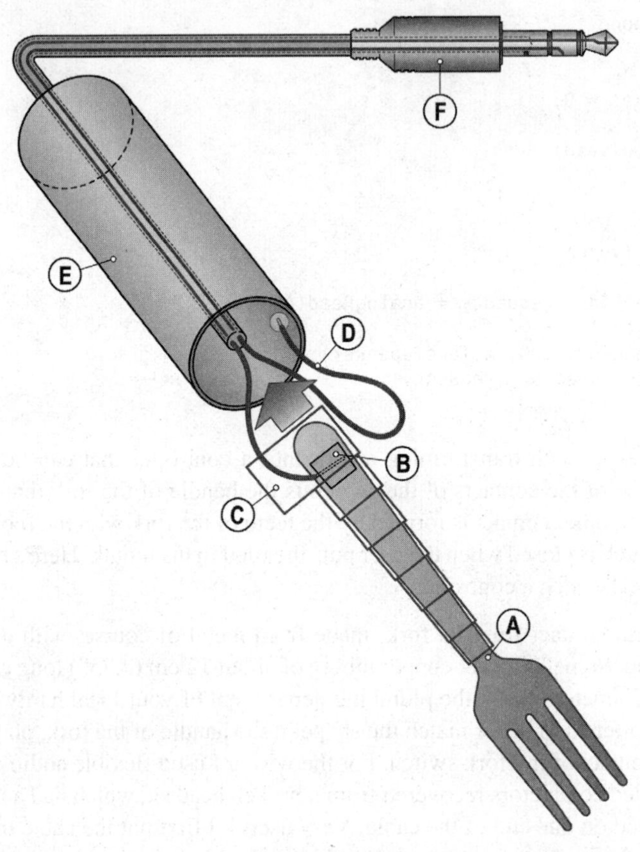

Figure 7-13 - Mechanical drawing of the Misophone. A: metal fork, B: self-adhesive copper tape + conductor 1, C: insulation tape, D: conductor 2, E: copper tube, F: 3.5 mm stereo audio jack.

of them to make your suffering during the meal bearable. You will see that after only a few misophonical dinners your friends will start declining your invitations and will stop inviting you to their homes, allowing you to spend all your free time again on programming microcontrollers.

Figure 7-14 - The Misophone and its shield built on an Arduino prototyping board.

7.5 A Bit of C++

From a C point of view the Arduino `tone` function is slightly special because it takes either two or three arguments. The API has other functions like this, tone is not alone. In C, it is possible to declare functions with an indefinite number of arguments (not to be confused with **`void`**), the best known example is probably `printf`. However, in the case of `tone` there is no question of an indefinite number of function arguments, but of a C++ function with an argument with a default value. If you specify a value for this argument, that will be the one that the compiler will use. If, on the other hand, you do not specify a value for this argument, the compiler will use a default value, in the same way as if you had specified it yourself. The default value is defined by the programmer, not the compiler, in the function declaration. Arguments with default values are always at the end of the argument list, it is not possible to follow such an argument by a "normal" argument. If the default value is not right for you, you must specify your own the value. Here's an example to clarify all this:

```
float foo(int age, int legs, int arms)
{
  return age / (legs + arms);
}
```

Suppose that two of the three arguments of the function `foo` have a default value. That can only be the arguments `legs` and `arms` because this type of arguments must be at the end of the argument list. If the default values (here both equal to 2) suit you, the function call can be done with a single argument (for age):

```
float useful_statistic = foo(61);
// useful_statistic = 15.25 (= 61/4)
```

If the default value for `arms` is fine, but, due to an unfortunate motor bike accident, not the value for `legs`, the function call requires two arguments:

```
float useful_statistic = foo(42,1);
// useful_statistic = 14 (= 42/3)
```

In all other cases, for example if `foo` is an octopus, you must call the function with three arguments, like this:

```
float useful_statistic = foo(5,0,8);
// useful_statistic = 0.625 (= 5/8)
```

When the default value for `legs` is fine, but not the default value for `arms`, the value for `legs` must be specified anyway, because the compiler is not able to guess that the second value is actually the third argument. Here's how to call `foo` in the case of Buddha Avalokiteshvara:

```
int useful_statistic = foo(1800,2,1000);[1]
// useful_statistic = 1.7964... (= 1800/1002)
```

With a little effort, you can do similar things in C, but it's not really worth it. It is easier to always specify optional argument(s).

In C++ there is another way of achieving similar things by defining several functions with the same name but with a different number of arguments and/or different types of arguments. A good example of such a function is the function `print` of the serial port (`Serial.print`), but there are more in other Arduino libraries. The function print accepts integers (`print(3)`), floating-point values (`print(3.14)`) and strings (`print("pi")`), but it treats them all differently. Overloading, which is how this technique is called, is not possible in C.

1. If `foo` has no arms and legs, what will happen?

7.6 The No in Arduino

The ADC of the AVR is a complex precision device that Arduino and its limited API do not really do justice. The MCU manufacturer has gone through a lot of trouble to provide a high-quality ADC with many options that are ignored by Arduino. Especially when it comes to noise suppression Arduino leaves a lot to be desired. The MCU's datasheet is pretty clear on this subject, and the detailed implementation instructions it provides are passed over by Arduino. Building a precision instrument based on an Arduino board is very difficult, even though the MCU is perfectly capable of precision measurements.

In addition to this, the analog part of the Arduino API does not provide functions for all analog devices built into the MCU, nor for all the options provided by the ADC. For instance, the Mega and Uno boards contain an analog comparator (AC in the datasheet) which is not supported by the API. On the Uno the comparator compares a voltage on pin 6 to a voltage either on pin 7 or on A0 to A5. Instead of pin 6, it is also possible to use the internal reference MCU. If the comparator is used with one of the inputs A0 to A5, the ADC cannot be active at the same time. The output of the analog comparator can generate interrupts or start the pulse capture function of timer/counter number one.

AD conversions can be triggered in several ways. Starting the ADC "manually" as is done by the API is one way of doing things, but triggering it by itself, a counter/timer, an external logic signal or via the integrated analog comparator is also possible.

The MCU on the Uno also contains a temperature sensor (on-chip) that the Arduino API does not know of. It is very easy to monitor the temperature of the chip, but not with the API. That is a shame, because to access the sensor only a small change to the `analogRead` function is needed (in the file `wiring_analog.c` in the folder <Arduino>\hardware\arduino\cores\arduino). Replace the '7' by an 'f' in the line

```
ADMUX = (analog_reference << 6) | (pin & 0x07);
```

like this:

```
ADMUX = (analog_reference << 6) | (pin & 0x0f);
```

and voila. Now you have a total of 16 (0 through 15) channels instead of 8 (0 through 7) and channel 8 corresponds to the on-chip temperature sensor. Selecting channel 14 allows you to measure the internal reference of 1.1 V. Channel 15 is connected to 0 V, which is convenient for calibrating a measurement or to verify the proper operation of the ADC, or, in some cases, of your system. The other channels are reserved and if you select one of them, the result will be unpredictable. Here is the sketch that tells you the temperature of the chip (after modifying the `analogRead` function):

```
/*
 * chip thermometer
*/

int A8 = 22;

void setup(void)
{
  Serial.begin(115200);
  analogReference(INTERNAL);
}

void loop(void)
{
  Serial.println(analogRead(A8));
  delay(1000);
}
```

This sketch only works with boards equipped with an ATmega48PA, an ATmega88PA, an ATmega168PA or an ATmega328PA. I defined the new analog input A8 as pin 22 so that `analogRead` will use it correctly. Once per second the sketch sends the measured temperature on the serial port. I did some experiments to determine the range of possible values. With my board immersed in a container filled with ice water, `analogRead` returned a value as low as 333 (358 mV). With the board in the kitchen oven set to 100 °C, the measured value went up to 444 (477 mV). Verifying the oven temperature with a thermocouple gave me 94 °C. According to the MCU's datasheet the sensor's sensitivity is 1 mV/°C, in my case its sensitivity seemed to be around 1.3 mV/°C, but this value is not very accurate.

7.7 Look Ma, No Arduino!

In the previous chapter we saw how to implement the digital inputs and outputs without using the functions provided by the Arduino API. Here we will do the same for the analog inputs.

The MCU's ADC consists of five 8-bit registers: ADMUX, ADCSRA, ADCSRB, ADCL and ADCH. A sixth register DIDR0 also exists (and even a seventh, DIDR2, on the Mega ATmega2560 board. For the curious among you, DIDR1 is used for the analog comparator) and allows the disconnection of the digital input buffer to save some energy.

The ADMUX register (ADC Multiplexer Selection Register) selects the analog input to connect to the ADC. The source of the reference voltage is selected here too and the register controls how the result of the conversion will be presented. ADCSRA and ADCSRB (ADC Control and Status Register A and B) manage the various ways to start a conversion, how interrupts will be generated and select the clock frequency used by the ADC. ADCL and ADCH (ADC Data Register) store the result of a conversion. It takes two 8-bit registers to do this as the ADC output is a 10-bit value.

Before you can start a conversion, you must configure the ADC: choose the input, the reference and the operation mode. Except maybe the choice of the input, the ADC configuration is often done only once, at the beginning of the program, but nothing prevents you to redo it differently at a later time. For a functionality similar to the one offered by the API, our reference shall be the supply voltage. The clock frequency determines the quality of the conversion, the slower the better. For a maximum resolution the clock frequency should be between 50 and 200 kHz. Beyond 200 kHz the low order bits become unusable due to noise, but a result of 8 bits, which is often enough, is still possible.

The conversion is started manually by setting the ADSC bit in ADCSRA. Then you must wait until the conversion completes before reading the result. Many MCUs, not just the AVR, use similar scenarios. Translated into C then we get this:

```
ADCSRA = 0x87;
ADCSRB = 0;
ADMUX = 0x40 | (input & 0x0f);
ADCSRA |= _BV(ADSC);
while (bit_is_set(ADCSRA,ADSC) != 0);
int result = ADCL;
result |= (ADCH << 8);
```

This piece of code converts the voltage present on `input` and puts the result in `result`. The first three lines configure the ADC and select the input you want. In the fourth line, the conversion is started by setting the ADSC bit in ADCSRA. The fifth line waits for the conversion to complete, which is the case when the ADSC bit gets cleared by the internal logic. The `bit_is_set` function that checks whether a bit is set or not is part of the Arduino compiler libraries, not of the API.

The 10-bit result is read in two times, first the lower eight bits are read and then the upper eight bits. The two characters "<<" correspond to a bit shift left operation, the 8 specifies the number of shifts. Shifting to the right is done with ">>".

Several techniques commonly used by MCUs are illustrated by this code fragment:

- The combination of several multi-bit variables using bitwise logical operators AND ('&') and OR ('|')[1] to make an 8-bit value (lines 1-3).The first line groups six parameters:

Bit name	Bit No.	Function	State
ADEN	7	Enable the ADC	Yes
ADSC	6	Start conversion	No
ADATE	5	Enable auto trigger	No
ADIF	4	Clear interrupt	No
ADIE	3	Enable ADC interrupts	No
ADPS	0 to 2	Divide clock by	128 (ADPS = 0x7)

The second line initializes two parameters in the ADCSRB register:

Bit name	Bit No.	Function	State
ACME	6	Enable analog comparator multiplexer	No
ADTS	0 to 2	Auto trigger source	Free running mode (ADTS = 0)

The other bits in this register are not used on our ATmega328.

The third line groups:

Bit name	Bit No.	Function	State
REFS	6, 7	Reference selection	AV_{CC} (REFS = 0x1)
ADLAR	5	Left adjust result	No
MUX	0 à 3	Input selection	Input No.

Bit 4 is not used.

1. Not to be confused with the Boolean operators AND ("&&") and OR ("||") used in logical conditions. & and | act on bits, && and || act on conditions.

- Start a transaction and wait until it completes (lines 4 and 5). Many functions that require multiple clock cycles to complete are initiated by setting a bit or writing a register. The end of the operation is flagged by a status bit or register that can be the same as the one that started the operation. Often it is also possible to trigger an interrupt (see Chapter 10) at the end of the operation so that the MCU can do something else instead of waiting for the operation to complete.

- Access registers in a specific order (lines 6 and 7). Here, reading the ADCL register first freezes the two data registers ADCL and ADCH until ADCH is read. This avoids overwriting the result by a new conversion that could be completed before ADCH or ADCL of the previous result are read. Often this method is used to automatically clear a register after reading it or to automatically start a function that requires an argument that occupies several registers.

It is essential to study the user manual of the MCU you are using to understand the functions of all the bits and registers and how to use them.

8. Communication: an Art and a Science

Over the past decades integrated circuit manufacturers, scientists, engineers and other developers of electronic devices have created a plethora of communication standards for all sorts of applications. If you look hard enough you can find a standard to solve almost every communication problem. This large amount of standards is quite understandable when you know that in even a simple microcontroller system communication exists at many levels:

- ✦ with you, the user;
- ✦ between systems;
- ✦ between components;
- ✦ between on-chip peripherals.

Communication with the user is through a human-machine interface (HMI) consisting of lights and LEDs, displays, buttons and other controls. As surprising as it may seem, this type of communication is hardly standardized, which is why you should read the user manual of every device you buy and not lose it; two things that few people are capable of. Human-machine communication is not the subject of this chapter (maybe of another book?).

In this chapter we will look at three levels of communication we can muster under the common denominator of communication between machines. A buzzword that we could use here is machine-to-machine (M2M) communication although in our case the machines are limited mostly to small electronic devices. Unlike human-machine communication, communication protocols for machines are often standardized and well documented. This unfortunately does not mean that we can ignore the manuals, because the way to specify a protocol varies from designer to designer and certain standards are extremely complicated.

You have probably noticed that I have started to mix the words communication and protocols. I realize this may be confusing, but I think I am in my right. My dictionary defines communication as "a process by which information is exchanged between individuals through a common system of symbols, signs, or behavior." That is pretty close to what I have in mind. The first half sounds like communication to me; the second half refers to the way to do it, i.e. the protocol. Communication is also the act of communicating. Adding an 's' we get communications: "the technology employed in transmitting messages", or "a system for transmitting or exchanging information" and "any of various professions involved with the transmission of information, such as advertising, broadcasting, or journalism".

Apart from the last, all definitions are fine for me for this chapter. They cover everything from the cable that connects the microcontroller with the computer, right up to the misunderstanding that makes us lose time.

Communication uses channels to transmit information. A channel can pass through a wired connection comprising of one or more conductors or over a wireless connection. The latter does not only mean radio waves; also think of communication by light (visible or not), by sound ("Shut up!"), by smell ("Hmm, do I smell cat pee?") or even by taste ("Yikes!"). The media that transports the information affects the way data can be exchanged, and the capacity and reliability of the channel determine in part the communication protocol needed to do it. The protocol encapsulates the data to be transmitted by adding additional information in order to ensure error-free reception. The more reliable a channel is, the simpler the protocol can be and the larger the amount of information that can be transmitted over it per unit of time.

The distance between two communicating devices is inversely proportional to the reliability of the channel. A cable is generally more reliable than a wireless connection, but it is not always practical or economical. The nature of the data to be transmitted also partly determines the complexity of the protocols to be used and the capabilities and properties of the channel required. An ATM and a bank do not communicate in the same way as a computer and a GPS receiver may do.

Multiple channels in parallel can carry more data per unit of time than a single channel[1]. There was a time when parallel communication was the standard in computer systems and data was carried by ribbon cables with dozens of conductors. Maybe your computer still has a parallel port? If so, you must have noticed the large size of the connector. It is precisely the size of cables and connectors that eventually pushed electronics manufacturers to develop communication systems with fewer conductors. You know the result: USB (Universal Serial Bus) and SATA (Serial Advanced Technology Attachment, used by computer disk drives) replaced flat cables and we started talking about serial communication. The manufacturers of microcontrollers, memories and peripherals have done the same and fewer and fewer chips have parallel ports[2]. There will always be some of course, but we will ignore them. In this book we are only interested in today's serial communication protocols that allow impressive speeds and that have nothing to envy to parallel communication protocols.

1. MIMO (Multiple In Multiple Out) techniques gave wireless systems the parallel port.
2. Inside the chip the peripherals typically communicate over a parallel bus.

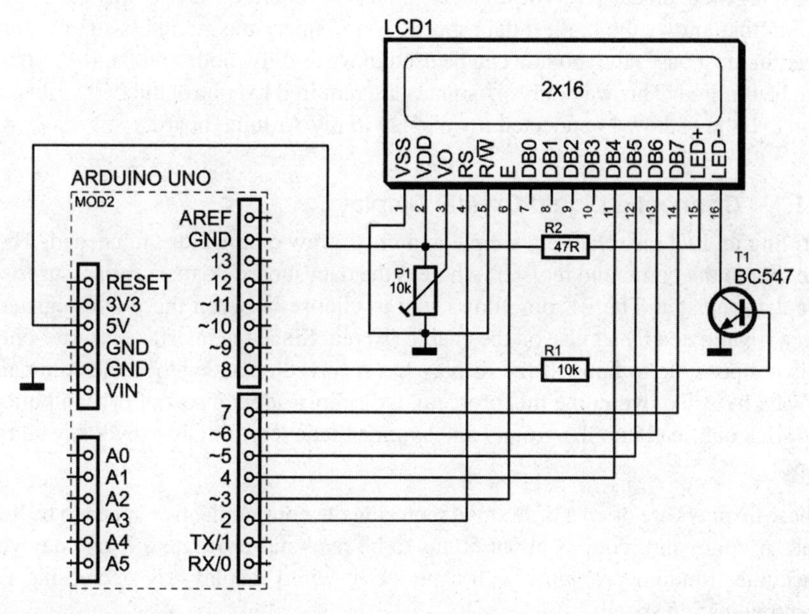

Figure 8-1 - A liquid crystal display connected in 4-bit mode. This arrangement works for all HD44780-compatible LCD modules. P1 adjusts the contrast, T1 controls the display's backlight.

8.1 Visualize Your Data

Before diving into serial communication, we will develop a tool to help us visualize data: a liquid crystal display (LCD). Ironically, this kind of module is usually controlled via a parallel port. Obviously there exist LCDs with serial interfaces, but the most common LCDs, and therefore most accessible to enthusiasts, are based on a chip from Hitachi, the HD44780 (or compatible). The number of lines (1 to 4) and the number of characters per line (8 to 40) depend on the model of the display. I will use an easy to find display with two lines of sixteen characters, but a type with four lines of twenty characters can also be used.

The HD44780 compatible LCDs have an 8-bit wide parallel interface, but as they are often used in applications where the number of ports is limited, the interface also has a 4-bit parallel mode where the eight bits of a byte are transferred in two steps. We can therefore speak of a hybrid interface: half parallel, half serial. The data bits are complemented by three control signals: enable (E), read/write (R/W) and a register select (RS). These three signals are required for two-way communication that allows the highest data speeds and to query the module's status. However, in most cases the module can be used in write-only mode and the R/W signal can be omitted. This way only six signals are required to control the LCD. Illustration 8-1 shows how I connected my display to my Arduino board.

8.1.1 Connect a Liquid Crystal Display

Driving an LCD is relatively simple and only a few commands are needed. They are sent to the command register, whereas the data should go to, surprise, surprise, the data register. The RS pin allows you to choose between these two registers. Data is validated by a pulse on the enable (E) pin. Since the interface is write-only, it is not possible to know if the display has received or understood the command or data byte. To overcome this problem, we simply adopt a sort of ostrich policy: we stick our head into the ground and assume that everything is okeydokey all the time.

These displays are often a little slow, requiring the communication speed to be limited. A command requires about 50 ms to be read and processed. If one day you encounter random problems with a perfectly wired display, try decreasing the communication speed.

Writing a driver for an HD44780-compatible display is not necessary because thousands of programmers have done this for you since the introduction of the chip. The internet has a ton of free drivers for this type of LCD, written in every programming language imaginable. Arduino also includes a library with all the functions needed to drive such a display and so we are not going to waste our time by repeating the exercise.

Note that this is an Arduino library and not a part of the Arduino API, meaning that accessing the library functions requires a (tiny) bit of extra work. Open a new sketch, in the menu click `Sketch`, then click `Import Library...` and select the library `LiquidCrystal` from the list that appears. The result of this operation is not spectacular, just a line of code inserted at the top of your sketch:

```
#include <LiquidCrystal.h>
```

This line informs the compiler that the file `LiquidCrystal.h` is now part of the sketch. The compiler will search all the files that make up a sketch for the definitions of the functions, constants and variables that it encounters in the program and now it will also look in `LiquidCrystal.h`. In this file the functions for controlling the LCD are declared. As you have imported this library, the next time you compile the sketch the file `LiquidCrystal.cpp` that contains the source code of this library will also be compiled.

`LiquidCrystal` is a flexible and easy to use library. It works with various types of displays, both in 8-bit and in 4-bit mode. First of all you should declare a `LiquidCrystal` object with as arguments the list of pins that are connected to the display. Then you should call the object's function `begin` to initialize the LCD. This function allows you to specify the number of rows and the number of characters per line. You can also specify a font size, but only if your display supports this option. Now you are in control of the LCD. Here's a little sketch that displays the text "hello everyone!":

```
/*
 * LCD
 */

#include <LiquidCrystal.h>

LiquidCrystal lcd(2,3,4,5,6,7);

void setup(void)
{
  lcd.begin(16,2);
  lcd.print("hello");
  lcd.setCursor(0,1);
  lcd.print("everyone!");
}

void loop(void)
{
}
```

The `lcd` object, declared at the top of the sketch, is connected to pins 2 to 7 in the order D4, D5, D6, D7, E, RS. Everything else is done in the function `setup`. Here the `lcd` object is configured as a display with two lines of sixteen characters, then the text "hello" is displayed using the print function of the lcd object. The second half of the message is displayed on the second line of the display after a call to `setCursor` which determines the position where character printing continues. Its second argument indicates the line of the display; the first argument indicates the position on the line. The first line is line 0, the first position is 0. A second call to `print` completes our friendly welcome message.

The library has other functions to perform more elaborate tasks like character scrolling, cursor control (visible or not, flashing or not), and creating custom characters, but for now we limit ourselves to these three functions.

Now that we have a working display with a partly parallel, partly serial interface we can move on to pure serial communication.

8.2 The Act of Communicating

Serial communication protocols can be divided into two groups: synchronous and asynchronous. The first uses two channels, one of which carries data, and the other a synchronization signal. Because this technique uses two channels, it is in principle limited to wired connections[1]. Asynchronous techniques combine the data and the synchronization signal in a single signal, so that one channel will suffice. Asynchronous communication can be accomplished with and without wires.

The AVR microcontroller found on the Arduino board is equipped with modules for both types of serial communication.

8.2.1 Asynchronous

Asynchronous communication is usually reserved for occasionally transmitting small data packets. That does not mean that an asynchronous communication bus cannot be used continuously or that it is impossible to send large packets, but this is less common.

Asynchronous communication links are everywhere thanks to the RS-232 standard developed in 1962. Many devices, instruments and other electronic apparatus have a so-called "serial port" and even if Intel and Microsoft have decided that the personal computer no longer needs one, this type of port remains extremely popular. Its ease of implementation – only two wires are necessary in most cases – and the availability of many serial communication programs for all operating systems have made the asynchronous serial port the pet port of the microcontroller system developer.

1. Synchronous communication can also refer to systems where nodes have the right to transmit only on predefined times. This technique is compatible with wireless connections.

Although it may seem paradoxical, asynchronous communication requires a synchronization mechanism. This mechanism operates at the level of bits and groups of bits or words. Two methods for bit synchronization can be distinguished:

- Fixing the duration of the bits and the time intervals between bits. If the receiver knows these times, it can reconstruct the message by sampling the signal at the right time. This technique is for example used by RS-232 and USB;
- Combining the synchronization signal and the data. Instead of signalling a bit by sending a high or low level, the bit is composed of two different levels. In this case the information is contained in the level changes, not in the levels themselves. If we ensure that each bit sent contains a level change, the receiver can reconstruct the clock and synchronize itself to the data flow. The popular Manchester coding uses this type of synchronization.

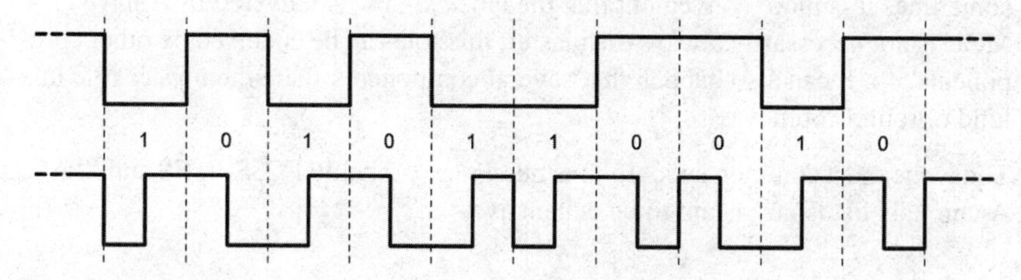

Figure 8-2 - An RS-232 signal (top) and a Manchester encoded signal (bottom). Note how the upper signal uses inverted logic, i.e. a '0' is represented by a high level, while a '1' is represented by a low level. In the bottom signal the bits are encoded in the level changes.

In addition to bit synchronization techniques, word synchronization methods are applied as well. In this case special bits identify the beginning and end of a word. For this to work, it is imperative that the receiver knows the length in bits of a word. The beginning of a word is indicated by a well-defined sequence of levels (start bits). Often the end of a word is also indicated by a sequence of predefined levels (stop bits). Usually a word contains eight bits, called a byte, but not always. Some industrial protocols for instance use words of nine bits.

The best known asynchronous bus is probably RS-232, but do not forget Ethernet, SATA, USB, CAN, LIN, 1-Wire and IrDA. This list is far from exhaustive.

8.2.2 Synchronous

Using two signals to transmit data synchronously is a good alternative to the parallel port. Indeed, this kind of bus (commonly called a bus because several transmitters and receivers can be connected to the same wires) offers high communication speeds and can transmit a large amount of data in a short time. The synchronous communication bus is often used to communicate with for example graphics displays, memories or audio and video codecs (a codec is a single device containing a (en)coder and decoder or ADC and DAC to convert an analog signal into a digital one and vice versa).

As noted above, a synchronous link requires two wires, one for the synchronization signal, the clock, and a second wire for the data. Synchronous communication systems are generally composed of a master and one or more slaves. The master decides on the direction of the data transfer (sending or receiving) and chooses the data destination or source. Often it also provides the clock, but not always, because sometimes it is more convenient that the clock signal is delivered by a slave. The MCU is not necessarily always the master; this role can be endorsed by other components too. It can also happen that several components share the master role in a kind of round-robin way.

Today the most common synchronous busses are probably I²C, SPI, I²S and JTAG. Again, this list is not meant to be exhaustive.

8.3 RS-232 or Serial Port?

The asynchronous communication standard EIA/TIA-232-E, introduced in 1962 and more commonly known as RS-232 (where RS stands for “Recommended Standard”), is extremely widespread. The suffix ‘E’ indicates that the standard has been revised four times since its appearance. The standard specifies a bidirectional serial connection of at least three wires: incoming data (RD), outbound data (TD) and a common reference (0 V). Additional signals for handshaking (RTS and CTS) and connection status (DSR, DTR and DCE) are also defined[1], but nowadays they are rarely used. RS-232 permits full-duplex communication, meaning that both ends of the link can transmit simultaneously. In reality, many devices only transmit or receive and one of the wires can be removed to create a two-wire connection.

1. In all, the EIA/TIA-232 standard defines 24 conductors, including the shield.

RS-232 is from the same epoch as TTL (Transistor-Transistor Logic) used in the first generation of microcontrollers, but despite this, their levels are not compatible. The voltage on the RD line should be between +3 V and +15 V to be recognized as a logic high level (5 V for TTL), and between −15 V and −3 V for a logic low level (0 V for TTL). A transmitter is expected to generate a voltage in the range of −15 V to −5 V for a logic low level and in the range of +5 V to +15 V for a logic high level. A bit disturbing here is the fact that a '0' (space) is indicated by a high level and a '1' (mark) by a low level. The standard also specifies a maximum rise time of 30 V/μs and a maximum communication speed of 20 Kbit/s in order to limit crosstalk between signals.

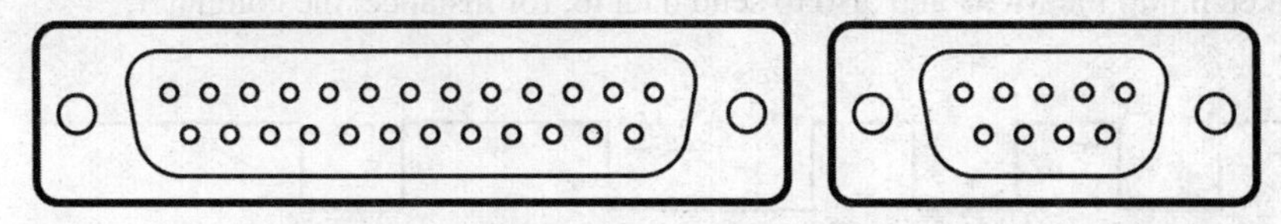

Figure 8-3 - D-subminiature connectors or sub-D. On the left a DB-25 model with 25 contacts, on the right a D***E***-9 with 9 contacts. The latter is often referred to as D***B***-9, which is incorrect, because the second letter refers to the shell size.

The connector is also standard: a real RS-232 port is equipped with a 25-pin sub-D connector. Given the size of this type of connector and the fact that most of the signals are used very little, a second connector type is commonly found, the 9-pin sub-D. Such a connector is still too bulky for miniaturized devices, and manufacturers tend to use a connector that works best for them regardless of the standard.

Today most RS-232-style connections are managed by microcontrollers and most of the RS-232 signals have become superfluous. MCUs operate at 5 V or less and communicate with similar systems nearby, therefore respecting the levels and polarities as specified by the standard is not really necessary. Communication speeds can go up to 1 Mbit/s or more. Since the standard is not respected (at all), it is better to speak of a "serial port" instead of an RS-232 port.

Inside a microcontroller it is the Universal Asynchronous Receiver-Transmitter (UART) module that handles this type of communication. It converts a stream of bytes into a stream of bits (we say serialize) and vice versa (parallelize). It also adds synchronization bits (start and stop bits) and the parity bit if necessary. This last bit provides a simple way to check if all the bits in a byte have been received correctly by ensuring that the number of ones in the packet is always even or odd. In practice this bit is rarely used. Sometimes the UART also provides the handshaking signals RTS, CTS, DTR and DSR (or more). However, in most cases they are not necessary, especially when other techniques are available to indicate that data has been received (such as interrupts that we will discuss in a later chapter).

The AVR's UART (or UARTs because the ATmega2560 on the Mega board has four of them) can work with packets that can contain 5 to 9 bits of data, an optional parity bit and one or two stop bits. Thus the length of a packet can vary from 7 up to 13 bits. The maximum communication speed is 2 Mbit/s and the UART does not offer additional RS-232 signals. In reality this is a USART where the 'S' stands for Synchronous, since the module is able to communicate asynchronously but also synchronously[1]. The USART plays an important role in Arduino; it is used to load a sketch into the MCU and also to send data to, for instance, the computer.

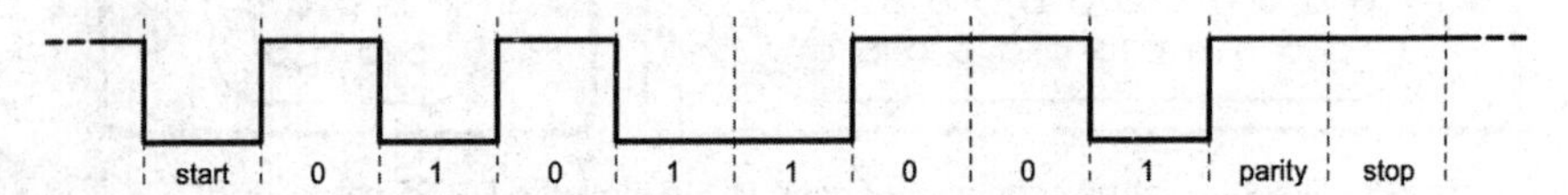

Figure 8-4 - An RS-232 packet consisting of a start bit, eight data bits, a parity bit (to make or keep the number of ones even) and a stop bit. If you look at the signal directly at the output of the MCU, you may see an inverted version of this signal. It is the RS-232 interface which inverts the bits.

After this long introduction you now probably want to get your hands dirty again. I propose to start simple. As you will see, Arduino is well equipped to implement asynchronous serial ports.

8.3.1 A Few Subtleties

In the previous chapters you have already seen how to get started with the serial port (I will drop the adjective asynchronous from now on): just put a statement `Serial.begin` in the function `setup`. The Mega board, or rather the MCU of the Mega board has four serial ports called `Serial` (without a 0), `Serial1`, `Serial2` and `Serial3`. Apart from their names, the way they work and the way to use them are identical.

A function `Serial.end` exists also, it frees the pins used by the serial port so you can use them for something else.

1. Arduino does not support the USART in synchronous mode.

The function `Serial.available` allows you to see if data bytes have been received. If this is the case, you can read them one by one with the function `Serial.read`.

Sending data sent over the serial port is done with the functions `Serial.write` or `Serial.print`. Both functions accept text strings as well as values, but there is a big difference: `Serial.write` sends raw data (often called "binary", even if that doesn't mean a lot), contrary to `Serial.print` that encodes the data to transmit as human-readable strings (in this case we often speak of ASCII, but that does not mean much either). For example, if you send the value 150 with `Serial.write`, the receiver will receive a single byte containing "1001 0110" (150 written as a binary number). If, on the other hand, you use `Serial.print`, the receiver will receive three bytes containing "0011 0001", "0011 0101" and "0011 0000", i.e. the characters '1', '5' and '0' encoded according to the ASCII character set. The following sketch illustrates this example. I used `Serial.println`[1] to make sure that the two output strings are shown on separate lines in the serial monitor.

```
/*
 * the difference between write and print
 */

void setup(void)
{
  Serial.begin(115200);
  Serial.println(150);
  Serial.write(150);
}

void loop(void)
{
}
```

If, after loading the sketch into the MCU, you open the IDE's serial monitor, you will see:

```
150
–
```

1. `println` is short for "print line". A line in this context is computer lingo for a carriage return (CR) followed by a line feed (LF).

The dash on the second line does not correspond to 150, but it is the serial monitor's rather inefficient way to display a byte that contains the value 150. Unfortunately, you have no way of knowing if this is supposed to be the dash ('-') or that it is a character that the serial monitor cannot display correctly.

The advantage of `Serial.print` is that the data displayed in a serial monitor will be readable, but this function cannot send all the values in the range 0 to 255. In many cases the readability of machine communication is of more importance than its speed or the number of characters to send and `Serial.print` is perfectly fine for the job. But if the goal is to send data as fast as possible or to send a byte that `Serial.print` cannot handle, then it is better to use `Serial.write`.

8.3.2 Chaining Characters

The Arduino API has grown over the years and today it contains several features that facilitate decoding strings received on the serial port. Communication protocols often use more or less fixed formats with well-defined synchronization fields or characters and the API provides functions to search or wait for such a field or character. Take for example the communication protocol NMEA 0183A as used by GPS receivers and other electronic devices of maritime origin. It is a protocol "in ASCII", meaning that it is based on text strings, so-called sentences, that are perfectly legible for a human being (legible is not a synonym for comprehensible). Here is a typical NMEA 0183A sentence as output by a GPS receiver (there are many others):

```
$GPGGA,064036.289,4836.5375,N,00740.9373,E,1,04,3.2,200.2,M,,,,
        0000*0E<CR><LF>
```

All valid sentences begin with a dollar sign ('$') and end with CR + LF (where CR means carriage return and where LF stands for line feed). The fields of the sentence are separated by commas. The header field following the dollar sign contains a two-character manufacturer or product code (GP is often used for a GPS receiver, but not always) followed by a three-character sentence identifier (GGA in this example) that lets you know what kind of information is contained in the sentence. GGA is a sentence that contains the position, the actual time, and some other information useful in navigation.

When you connect a GPS receiver to your Arduino board[1] and then launch a sketch to decode the GPS data, the sketch is not synchronized with the sentence. It has to wait until a dollar sign is received before starting to decode the incoming data. This is not very difficult to program, but the function `Serial.find` (new in Arduino

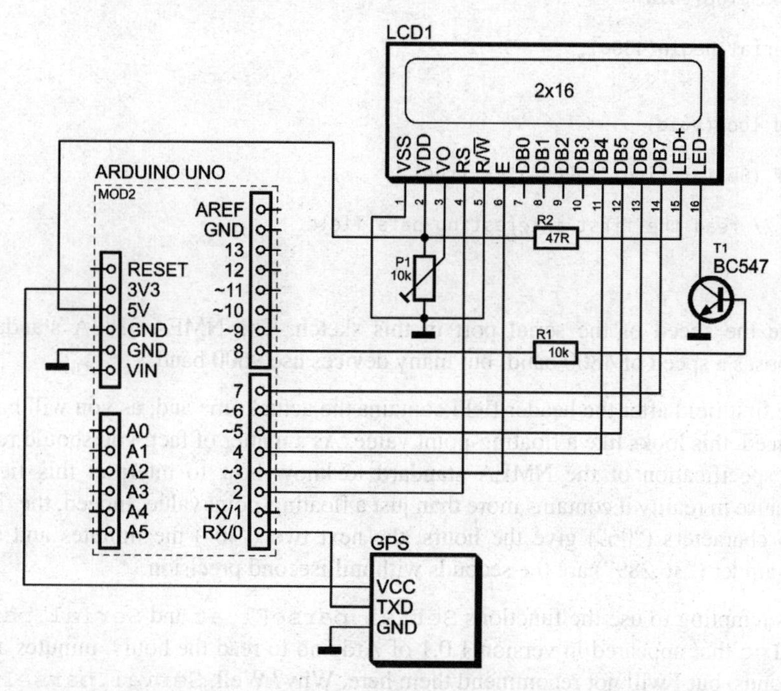

Figure 8-5 - For the experiments that follow we need a GPS receiver. Here is how to connect a module that is powered by 3.3 V. For a 5 V module, connect VCC to 5V.

version 1.0.1) can do it for you. Furthermore, if you plan to always use the same GPS receiver of which only the GGA sentence is of interest, you can even wait until you receive the part "$GPGGA," (including the comma). This way you will know that when `Serial.find` says I'm good, I saw "$GPGGA,", you can read the first field with interesting data without further ado.

```
/*
 * Serial.find
 */
```

1. If your GPS needs to be powered from 3.3 V you can connect it anyway, because the MCU on the Arduino board has no problems understanding 3.3 V signals. What's more, the board has a 3.3 V voltage regulator. It is generally in the other direction, from the 5 V MCU to a 3.3 V peripheral, that signal level differences may pose problems and thus require the use of level adapters.

```
void setup(void)
{
  Serial.begin(4800);
}

void loop(void)
{
  if (Serial.find("$GPGGA,")==true)
  {
    // read the first interesting data field.
  }
}
```

Note the speed of the serial port in this sketch. The NMEA 0183A standard imposes a speed of 4800 baud, but many devices use 9600 baud.

The first field after the header field contains the actual time and, as you will have noticed, this looks like a floating-point value. As a matter of fact, you should read the specification of the NMEA standard to know how to interpret this field, because in reality it contains more than just a floating-point value. Indeed, the first two characters ("06") give the hours, the next two ("40") the minutes and the remainder ("36.289") are the seconds with millisecond precision.

It is tempting to use the functions `Serial.parseFloat` and `Serial.parseInt` that appeared in version 1.0.1 of Arduino to read the hours, minutes and seconds, but I will not recommend them here. Why? Well, `Serial.parseInt` is unusable because it will read the characters "064036" up to the decimal point and then return an integer value of 64,036. `Serial.parseFloat` will return the value 64,036.289 and you still will have no idea what time it is. Other fields of the GGA sentence also contain groups of data or even letters, so how do you process such fields?

One method is to read the characters one by one from the serial port and decode them on the fly. Personally I find it more practical to wait for the reception of an entire sentence and then interpret the fields, as follows:

```
/*
 * read nmea sentence
 */

#include <LiquidCrystal.h>
#define LCD_LINES 2
#define LCD_CHARACTERS_PER_LINE 16
LiquidCrystal lcd(2,3,4,5,6,7);

#define LF 0x0a
#define CR 0x0d
```

```
#define SENTENCE_MAX 128
char sentence[SENTENCE_MAX];
int sentence_length;

#define WAITING 0
#define COLLECTING 1
#define DONE 2
int state;

void lcd_write_line(int line, char str[])
{
  lcd.setCursor(0,line);
  int length = min(strlen(str),LCD_CHARACTERS_PER_LINE);
  for (int i=0; i<length; i++)
  {
    lcd.print(str[i]);
  }
}

void setup(void)
{
  lcd.begin(LCD_CHARACTERS_PER_LINE,LCD_LINES);
  Serial.begin(4800);
  state = WAITING;
  sentence_length = 0;
}

void loop(void)
{
  if (Serial.available()>0)
  {
    char ch = Serial.read();
    if (ch=='$')
    {
      state = COLLECTING;
      sentence_length = 0;
    }
    else if (ch==CR || ch==LF)
    {
      state = DONE;
      sentence[sentence_length] = 0;
    }
    if (state==COLLECTING && sentence_length<SENTENCE_MAX)
    {
      sentence[sentence_length] = ch;
      sentence_length++;
    }
  }

  if (state==DONE)
  {
```

```
    // display a maximum of 16 characters on the first line
    lcd_write_line(0,sentence);
    if (strlen(sentence)>LCD_CHARACTERS_PER_LINE)
    {
      // if there are more characters,
      // display a maximum of 16 characters on the second line
        lcd_write_line(1,sentence+ LCD_CHARACTERS_PER_LINE);
    }
    state = WAITING;
  }
}
```

In this sketch I dropped the function `Serial.find` to find the beginning of the NMEA sentence, because I do not find it flexible enough. My approach allows us to read every NMEA sentence, not just those that begin with "$GPGGA,". Besides, if you execute this sketch with a real GPS receiver, it is likely that you do not even see the GGA sentence. The reason is that a GPS receiver typically sends a bunch of different sentence types once per second. This sketch shows them all, but since the sentences arrive very quickly one after the other, you will only be able to see the last one. All you will see is a stream of characters rapidly passing by and ending with the display of the last sentence. If the GGA sentence is not the last one, you will not see it. Later on I will show you how to filter the data in order to retain only the interesting sentences.

The way the sketch works is not very difficult to understand, because it is based on a state machine. The one used here has three states: `WAITING`, `COLLECTING` and `DONE`. At start-up, the sketch enters the `WAITING` state in which it remains until the arrival of a '$' character. This happy event signals the beginning of a sentence and the sketch moves to the `COLLECTING` state. The sentence length is set to zero and the dollar sign is stored in the sentence buffer. While in the `COLLECTING` state and as long as there is space in the sentence buffer, all received characters will now be recorded, except for the CR and LF characters. The reception of one of these two characters causes the state to change to the last state, `DONE`. A null character is stored in the sentence buffer to make it a standard zero-terminated C string (see Section 6.14, page 108). The change to the `DONE` state is detected at the end of the function loop and a maximum of 32 characters of the received sentence is displayed on the two lines of the LCD. Once the sentence has been printed, the state machine returns to its initial `WAITING` state and the program starts looking out for the next dollar sign.

8.3.3 Breaking the Chains

The previous sketch introduced two new elements. The first is the way to select the characters that will be displayed on the second line of the LCD. If the sentence contains more than sixteen characters (in this case, LCD_CHARACTERS_PER_LINE for a more generic case), the first sixteen are displayed on the first line and the remaining characters – but not more than sixteen – are displayed on the second line. To do so it is necessary to indicate the start of the second batch of sixteen characters, and in C this can be done simply by adding the offset of the first character to display to the sentence buffer. This may seem strange, but it is actually quite normal and even defendable. Here we touch upon a feature of C hated by some and adored by others: pointers. I will explain the concept of pointers at the end of this chapter[1], because I do not want you to lose the thread of my explanation. For now just accept that in C you can do things like this.

The second novelty of the sketch is the use of the function strlen contained in one of the core libraries of the C compiler. strlen is short for string length and the function calculates the length of a zero-terminated string. This function is part of a family of functions whose names all begin with str (strcpy, strcmp, strcat, strstr, etc.) and that operate on character strings terminated by a null character (again, see Section 6.14, page 108). The Arduino API includes objects of the type String (with a capital) that facilitate the use of these str functions. These objects are also used to circumvent the thorny problem of pointers. Here I wanted to show you what to do when you do not have access to String objects, unfortunately a rather common situation. Next is a sketch with the same functionality as the previous one, but using a String object to store the NMEA sentence.

```
/*
 * read nmea sentence 2
 */

#include <LiquidCrystal.h>
#define LCD_LINES 2
#define LCD_CHARACTERS_PER_LINE 16
LiquidCrystal lcd(2,3,4,5,6,7);

#define LF 0x0a
#define CR 0x0d
String sentence;

#define WAITING 0
#define COLLECTING 1
#define DONE 2
int state;
```

1. This phrase is a pointer.

```
void lcd_write_line(int line, String str)
{
  lcd.setCursor(0,line);
  for (int i=0; i<min(str.length(),LCD_CHARACTERS_PER_LINE); i++)
  {
    lcd.print(str[i]);
  }
}

void setup(void)
{
  lcd.begin(LCD_CHARACTERS_PER_LINE,LCD_LINES);
  Serial.begin(4800);
  state = WAITING;
}

void loop(void)
{
  if (Serial.available()>0)
  {
    char ch = Serial.read();
    if (ch=='$')
    {
      state = COLLECTING;
      sentence = "";
    }
    else if (ch==CR || ch==LF)
    {
      state = DONE;
    }
    if (state==COLLECTING)
    {
      sentence += ch;
    }
  }

  if (state==DONE)
  {
    // display a maximum of 16 characters on the first line
    lcd_write_line(0,sentence);
    if (sentence.length()>LCD_CHARACTERS_PER_LINE)
    {
      // if there are more characters,
      // display a maximum of 16 characters on the second line
      lcd_write_line(1,sentence.substring(LCD_CHARACTERS_PER_LINE));
    }
    state = WAITING;
  }
}
```

Note how the function `strlen` has been replaced by the function `sentence.length` and how the use of `sentence.substring` avoids the addition of an offset value to `sentence` to display the characters on the second line of the LCD. This sketch is a little shorter than the previous one, because it is not necessary to manually count the number of characters received, the `String` object does it for you. However, when you compile this example, you will notice that the executable is much larger than the one produced for the first sketch: 5,434 bytes against 3,492 bytes (on my computer). You see that comfort has a price.

8.3.4 An NMEA 0183A Decoder

Now that you know how to receive NMEA data, we can extend the sketch to display the geographical location sent to us by the GPS receiver. The coordinates of this location are contained in the four fields following the time field, two fields for the latitude and two for the longitude. In our example sentence the latitude is "4836.5375, N" which corresponds to 48 degrees and 36.5375 minutes of arc North. The longitude "00740.9373, E" is 7 degrees and 40.9373 minutes of arc East. As you can see, the formats of these two coordinate parts are not quite identical. The degrees of the longitude are coded in three digits, 0 to 180°, while those of the latitude occupy only two digits, from 0 to 90°. Furthermore, the field length is variable because the number of decimal places is not fixed by the NMEA 0183A standard. Therefore we cannot count characters to go from field to field; instead we need a function that allows us to recognize the beginning of a field by detecting the commas separating the fields. The `String` object has a function `indexOf` that can help us to do this. The function `substring` of the same object will allow us to extract the degrees and minutes of arc.

```
/*
 * read nmea sentence 3
 */

#include <LiquidCrystal.h>
#define LCD_LINES 2
#define LCD_CHARACTERS_PER_LINE 16
LiquidCrystal lcd(2,3,4,5,6,7);

#define LF 0x0a
#define CR 0x0d
String sentence;

#define WAITING 0
#define COLLECTING 1
#define DONE 2
```

```
int state;

int nmea_find_field(int nr, String str)
{
  if (nr==0) return 0;
  int i = 0;
  do
  {
    i = str.indexOf(',',i);
    if (i>=0)
    {
      nr--;
      i++;
    }
  }
  while (nr>0 && i>0);
  return i;
}

void lcd_write_line(int line, String str)
{
  lcd.setCursor(0,line);
  for (int i=0; i<min(str.length(),LCD_CHARACTERS_PER_LINE); i++)
  {
    lcd.print(str[i]);
  }
}

void setup(void)
{
  lcd.begin(LCD_CHARACTERS_PER_LINE,LCD_LINES);
  Serial.begin(9600);
  state = WAITING;
}

void loop(void)
{
  if (Serial.available()>0)
  {
    char ch = Serial.read();
    if (ch=='$')
    {
      state = COLLECTING;
      sentence = "";
    }
    else if (ch==CR || ch==LF)
    {
      state = DONE;
    }
    if (state==COLLECTING)
    {
```

```
      sentence += ch;
    }
  }

  if (state==DONE)
  {
    if (sentence.startsWith("$GPGGA")==true)
    {
      int lat_start = nmea_find_field(2,sentence);
      int lon_start = nmea_find_field(4,sentence);
      int lon_end = nmea_find_field(6,sentence);
      lcd_write_line(0,sentence.substring(lat_start,lon_start-1));
      lcd_write_line(1,sentence.substring(lon_start,lon_end-1));
    }
    state = WAITING;
  }
}
```

This sketch is very similar to the previous sketch; I even hesitated to incorporate it in full in the text. The differences are at the end of the function `loop`, in the detecting portion of the `DONE` state, and the addition of a new function `nmea_find_field`. This function allows you to find the beginning of a field; it returns the index of the character following the comma of the requested field. If the field does not exist, the function returns –1. Note that the fields are numbered from zero to n-1. If you ask for field zero, you will receive an index of zero.

When the `DONE` state is detected in the function `loop`, the sketch checks if the received sentence is of the GGA kind using the function `startsWith`. If this is the case, we can extract the latitude and longitude of the geographical position calculated by the GPS receiver. Since this information is always available in the same fields, we can use hard-coded values in the calls of `nmea_find_field`. Note that the latitude and longitude each occupy two fields, one for the numerical value and one for the cardinal points North, East, South or West. The function `substring` performs the actual data extraction using the field indices obtained by `nmea_find_field`. The results are then displayed on the LCD, the latitude on the first line and the longitude on the second.

With this kind of work, always be very careful with the beginning and the end of the string to extract. Often a character is forgotten or, on the other hand, a character too many is extracted. Here for example `substring` extracts the subsentence from `lat_start` to `lon_start-1`. But `lon_start` contains the index of the character following the comma that indicates the start of field number four; `lon_start-1` therefore corresponds to the index of the comma itself.

Fortunately, the function substring stops just before this value and the comma will not be part of the subsentence of interest. Phew, that was close. Always read the description of a function to ensure that it stops there where you expect it to stop.

Now you have all the elements needed to retrieve the information you want from any NMEA sentence and to display it. I suggest you do some experiments such as displaying the time or the speed to familiarize yourself with string operations. The GPS receiver is an interesting tool for this because it sends lots of information.

8.3.5 Mutatis Mutandis

The next step is to convert "ASCII" data into numerical "binary" values that can be used in calculations. This is a very common task in programming, and the C language libraries offer some easy-to-implement functions to achieve this, like atoi (ASCII to integer, for converting a character string into an integer value), atol (ASCII to long, converts a string to an integer XL) or atof (ASCII to float, to convert a string to a floating-point value). The Arduino API provides similar functions for Stream objects and String objects. Our NMEA sentence is a String object so we can use the function toInt to convert a string to an integer. However, the String object does not have a function to convert a string into a floating-point value. This is not really a problem because we can easily write one ourselves. First we convert the integer part (the part before the decimal point), then the part after the decimal point and finally we add the two parts together without forgetting to divide the second part by a power of ten that corresponds to the number of decimal places. Here is a function that implements this technique:

```
float string_to_float(String str)
{
  int dp = str.indexOf('.');
  decimal_digits = str.length() – dp - 1;
  float precision = pow(10,decimal_digits);
  float result = str.substring(0,dp).toInt();
  result += str.substring(dp+1).toInt()/precision;
  return result;
}
```

First, the position of the decimal point is determined. The number of decimal places, obtained by subtracting the position of the decimal point from the length of the character string, is used to calculate the precision. The number of decimal places is stored in the global variable decimal_digits to be reused elsewhere. Then the digits in front of the decimal point are converted to an integer (but stored in a **float**), and, on the next line, the decimals are converted before being divided by precision and added to the integer part.

With this function we can finish the conversion of the latitude and the longitude values. It is possible to write a function that can handle both values. Here is what I came up with:

```
float string_to_lat_lon(String str)
{
  int dp = str.indexOf('.');
  int cardinal = str.indexOf(',');
  float degrees = str.substring(0,dp-2).toInt();
  degrees += string_to_float(str.substring(dp-2,cardinal))/60.0;
  if (str[cardinal+1]=='S' || str[cardinal+1]=='W') degrees *= -1.0;
  return degrees;
}
```

The main difficulty is to determine the number of digits used to encode the degrees. We can find this number by starting at the decimal point and then go two digits backwards (to the beginning of the string) because we know that the minutes of arc always occupy two digits. The remaining digits that separate us now from the beginning of the string are necessarily used by the degrees.

A second obstacle is the field that contains the cardinal point ('N', 'E', 'S' or 'W') and which must not be converted. Here I could have used the `nmea_find_field` function, but since there will always be only one comma in `str`, I preferred to use `indexOf` because it is slightly faster.

Once we know where the data is, converting it into a numerical value is a breeze. It starts with the conversion of the degrees into an integer that we store in a variable of type **`float`**. The minutes of arc are converted using the function `string_to_float`, of which the result is divided by 60 (remember, one degree is divided in 60 minutes of arc, not 100) and added to the degrees. For the cardinal points south ('S') and west ('W') the degrees must be multiplied by –1 to maintain compliance with the standards used in navigation.

Have you noticed how in the previous two functions I put function calls "in series" with a dot? For example, `str.substring(0,dp-2).toInt()`. This is possible because str is an object of type `String` and the function `substring` returns also an object of type `String`. Objects can have functions and variables that can be used by placing a dot after the object's name followed by the name of the function or the variable. The notion of objects is derived from C++. Since the Arduino API is a mix of C and C++, C++ constructions are freely mixed with constructions typical for C, and it's up to you to know whether a variable is (part of) an object or not. This makes sketches inelegant[1].

1. Yes, elegance can be a measure of quality for a program.

The following function retrieves the latitude or longitude as a floating point value from a GGA sentence:

```
float get_lat_lon(int field, String sentence)
{
  int value_start = nmea_find_field(field,sentence);
  int value_end = nmea_find_field(field+2,sentence);
  return string_to_lat_lon(sentence.substring(value_start,value_end-1));
}
```

8.3.6 Make a U-turn Now

With our new `get_lat_lon` function we can make a funny little application: an automotive navigation system confuser. That's right. If you insert an Arduino board between the GPS receiver and the device that displays the geographical position of the vehicle overlaid on the map, you can modify the displayed position and you can influence the information provided by the navigation system. "Make a U-turn now", "Turn left here", etc. you can now trigger these messages whenever you want. How?

Actually it is quite simple. For example, to trigger the message "Make a U-turn now" simply invert the direction calculated by the car navigation system. This is possible if you intercept the GGA sentences and modify them before you send them to the display. At any moment, after having recorded a starting point, you can invert the travel direction with respect to this point. The navigation system will think that the car is moving in the wrong direction and it will start saying "Make a U-turn now". After some thirty seconds or so you can stop modifying the sentences (to reactivate it a few minutes later of course). You can also wait for the driver to turn around first before stopping to tamper with the GGA sentences. That way you will trigger a new, but this time justified “Make a U-turn now” message which will confuse the driver even more. With such a car navigation system confuser you will notice that your friends and family no longer want to travel with you, allowing you to spend all your free time again on programing microcontrollers. So how do we go about?

Creating the car navigation system confuser is now relatively easy since we already have almost all the necessary elements to do so. The only feature missing is the one that allows us to alter the geographical position in a GGA sentence. Performing this task would have been simple if only the `String` object had the required functions (or methods, as the C++ expert says). Indeed, `String` offers a nice function `substring` to extract a subsentence, but the function to do the inverse, i.e. inserting a subsentence, does not exist[1] and so we must write it ourselves. We will also need the inverse of the function `string_to_lat_lon` to convert latitude and longitude values to a string. Finally, the `String` object does not have a method for writing a floating-point value; again it is up to us to provide one. In the end, we need quite a lot of small functions. It is possible to extend the `String` object with our own functions, but that will break the compatibility with future versions of Arduino.

```
void string_set_substring(int start, String &dst, String src)
{
  for (unsigned int i=0; i<src.length(); i++)
  {
    dst[start+i] = src[i];
  }
}

void int_to_string_zeropad(long value, int digits, String& s)
{
  while (value<pow(10,digits-1) && digits>1)
  {
    s += '0';
    digits -= 1;
  }
  s += value;
}
```

1. It would have been possible to use the function `replace` to re-find the extracted subsentence and replace it with the new subsentence, but that would not have been very elegant.

```
void float_to_string(float f, int digits, int decimals, String &s)
{
  long fi = (long)f;
  f = abs((f-fi)*pow(10,decimals));
  long fd = (long)f;
  int_to_string_zeropad(fi,digits,s);
  s += '.';
  int_to_string_zeropad(fd,decimals,s);
}

void lat_lon_to_string(float value, boolean is_longitude, String &s)
{
  boolean negative = value<0;
  value = abs(value);
  int degrees = (int)value;
  int_to_string_zeropad(degrees,is_longitude?3:2,s);
  float_to_string((value-degrees)*60,2,decimal_digits,s);
  s += ',';
  if (is_longitude==true)
  {
    if (negative==true) s += 'W';
    else s += 'E';
  }
  else
  {
    if (negative==true) s += 'S';
    else s += 'N';
  }
}

void set_lat_lon(int field, float value, boolean is_longitude,
                                                      String &dst)
{
  String str;
  int pos_start = nmea_find_field(field,dst);
  lat_lon_to_string(value,is_longitude,str);
  string_set_substring(pos_start,dst,str);
}

void set_gga_position(float latitude, float longitude, String &gga)
{
  set_lat_lon(2,latitude,false,gga);
  set_lat_lon(4,longitude,true,gga);
}
```

Let's start reading at the end, in the function `set_gga_position`. This function is called with the latitude and longitude as arguments in the form of floating-point values and the GGA target sentence stored in a `String` object. From here we call `set_lat_lon` twice, once for the latitude and once for the longitude. In `set_lat_lon` we first search for the position in the target sentence where we will write the latitude or the longitude. Then the value is converted to a `String` object with the function `lat_lon_to_string`, and then it is copied to the right place in the target sentence using `string_set_substring`.

The `lat_lon_to_string` function first extracts the sign of the value of the latitude or longitude because we need it later, the value itself is written in the sentence without a sign. Note how the result of a comparison is used here directly to detect the sign. In C, false usually equals to zero and true generally equals to one (good programmers never rely on such assumptions).

We should pad the degrees field with zeros when the printed degrees do not occupy all the digits reserved for them. For example, if the longitude is 2, you have to write "002". The latitude degrees occupy a maximum of two positions. Filling a field like this is called zero padding and it is the `int_to_string_zeropad` function who takes care of this.

The rest of the function is almost entirely devoted to the processing of the cardinal points. Remember that the global variable `decimal_digits` has been set when the function `string_to_float` was called during the decoding of the GGA sentence. It may not be very elegant, but it is defendable, because this variable is a kind of system parameter.

Converting a **`float`** to a `String` is achieved by the `float_to_string` function. It is the counterpart of the function `string_to_float`; it cuts the floating-point value into an integer part (before the decimal point) and a fractional part (after the decimal point). Both parts are integers and both must be written into the string using zero padding, it is again `int_to_string_zeropad` who takes care of this once more.

We conclude with the function `string_set_substring` that writes a string at the desired position in a longer string.

But that's not all. No sir! There is also a checksum to update. A what? A checksum. Let me explain. Due to noise and interference it can happen that the receiver interprets one or more bits of a sentence incorrectly making the received data invalid. To detect (not prevent) this type of error the transmitter can add a field to the sentence that allows the verification of the data contained in the sentence. This field, called the checksum, is calculated over all the data contained in the sentence. The receiver performs the same calculation on the received data and if it does not find the same result as the transmitter, it rejects the data. NMEA sentences contain such a checksum and since we changed the data in the sentence, we must recalculate the checksum's value. If this value is incorrect, the car navigation system will reject our modified sentences and our joke will fall flat. Here is how you can calculate the checksum:

```
const char to_hex[16] =
{
  '0', '1', '2', '3', '4', '5', '6', '7',
  '8', '9', 'A', 'B', 'C', 'D', 'E', 'F'
};

void nmea_calculate_checksum(String& str)
{
  unsigned int i;
  unsigned char checksum = 0;
  for (i=1; i<str.length(); i++)
  {
    if (str[i]=='*') break;
    checksum ^= str[i];
  }
  checksum &= 0xff;
  str[i+1] = to_hex[(checksum&0xf0)>>4];
  str[i+2] = to_hex[checksum&0x0f];
}
```

The sum is calculated by taking the exclusive or (XOR) of all bytes between '$' and '*', without including these two characters. In C the XOR operation is represented by a hat (or "caret") character ('^'). The resulting byte can have a value between 0 and 255 and it is added to the sentence as a string of two hexadecimal characters, one for each 4-bit nibble. A little array, `to_hex`, can be used to easily convert a nibble to an ASCII character; converting a hexadecimal byte therefore requires two consultations of this table.

Note that many devices supposedly compatible with the NMEA 0183A standard incorrectly calculate the checksum. It is for this reason that some equipment or software that can handle NMEA data has an option to disable the checksum verification.

For completeness sake, here are the functions `setup` and `loop` and the global variables and definitions of the sketch of the car navigation system confuser:

```
/*
 * gps inverter
 */

#define LF 0x0a
#define CR 0x0d
String sentence;

#define WAITING 0
#define COLLECTING 1
#define DONE 2
```

```
int state;

float fix_lat;
float fix_lon;
int decimal_digits;
boolean inverting = true;
boolean need_fix = true;
int seconds = 0;

void setup(void)
{
  Serial.begin(9600);
  state = WAITING;
}

void loop(void)
{
  if (Serial.available()>0)
  {
    char ch = Serial.read();
    if (ch=='$')
    {
      state = COLLECTING;
      sentence = "";
    }
    else if (ch==CR || ch==LF)
    {
      state = DONE;
    }
    if (state==COLLECTING)
    {
      sentence += ch;
    }
  }

  if (state==DONE)
  {
    if (sentence.startsWith("$GPGGA")==true)
    {
      float latitude = get_lat_lon(2,sentence);
      float longitude = get_lat_lon(4,sentence);
      if (inverting==true)
      {
        if (need_fix==true)
        {
          fix_lat = latitude;
          fix_lon = longitude;
          need_fix = false;
          seconds = 0;
        }
        else
```

```
        {
          float delta_lat = fix_lat - latitude;
          float delta_lon = fix_lon - longitude;
          latitude = fix_lat + delta_lat;
          longitude = fix_lon + delta_lon;
          set_gga_position(latitude,longitude,sentence);
          nmea_calculate_checksum(sentence);
          seconds += 1;
          if (seconds>30)
          {
            inverting = false;
            need_fix = true;
          }
        }
      }
    }
    state = WAITING;
    Serial.println(sentence);
  }
}
```

The function `setup` is so simple that it is hardly worth talking about. In `loop` we find the same code for receiving the NMEA sentences as in the previous examples. Things only get interesting in the second part of `loop`. If the confuser is idle, the received sentence is not modified and it is simply forwarded to the serial port. If, on the other hand, the confuser has been enabled (`inverting` equals `true`) and it has just received a GGA-type sentence, the geographical position is extracted. The first time after activation of the confuser we must store this position as a reference or as a center of symmetry. The seconds counter must also be reset. The next time we pass through `loop` the opposite position of the current position with respect to the center of symmetry is calculated and written to the GGA sentence. The checksum is corrected, the seconds counter is updated and the modified sentence is sent to the serial port. After 30 seconds the confuser is turned off automatically.

As you may have noticed, there is no mechanism to trigger the confuser. The variable `inverting` is never set to `true` (except during the initialization of the sketch). This is done on purpose, because it is up to you, the prankster, to decide the right time to activate the confuser. As with all pranks, timing is primordial, so you will have to decide the right moment to launch it, maybe with a pushbutton, maybe after a specified interval or maybe at a specific location.

8.3.7 A Curly Symbol

You probably have noticed ampersands ('&') in front of some arguments in the definition of certain functions. This symbol signifies that these arguments can be changed in the function in question, and that the changes will be retained when the function is terminated. This is another way to return a result. Refer to Section 8.8 (page 214) on pointers at the end of this chapter to learn more about this technique.

Thus ends our long journey through the Kingdom of the Asynchronous Serial Port. Since this type of port is often used to transmit or receive ASCII strings, I have taken the opportunity to explain how to work with this type of data. Every real programmer, including you, has to know how to work with text strings. Now let's move on to the other types of serial communication.

8.4 Two-Wire Connections

Twi (or Akan or Fante) is a language spoken by about seven million people in Ghana. Surprisingly, many microcontrollers also speak TWI and the AVR that we are beginning to become familiar with is no exception. The TWI spoken by MCUs is not a dialect of the Ghanaian Twi, but of I²C. In fact, dialect is a big word, as TWI and I²C are so similar that the only difference is in the name. TWI (Two-Wire Interface) was probably created to avoid legal or financial problems or both relating to the use of the acronym I²C of which Philips was the owner.

The I²C bus (from Inter Integrated Circuits) was developed by Philips in the 1980s (NXP now manages this standard) to connect a microprocessor or microcontroller in an easy manner to other integrated circuits in – at the time of invention – mostly television sets made by the brand and its sub-brands. Little by little the bus became popular and many I²C-compatible components have seen the light. I²C is a bus where a master device communicates with one or more slave devices and where roles may change along the way. There are several variants of the bus and they offer different communication speeds. The standard was revised not later than 2012 (revision 4.0) when the Ultra Fast-mode or UFm was added that allows communication speeds up to 5 MHz. SMBus (and PMbus) is a variant of the I²C bus and some MCUs handle both protocols natively. Because of the additions and evolutions the I²C modules are becoming more and more complex and it is not obvious to write a driver that can handle all modes. To overcome this, certain recent MCUs not only support the bus with hardware, they also integrate the I²C driver software to use it.

If we do not count the reference signal or common, the I²C bus is a two-wire link with a data (SDA) and a clock (SCK) line. The nodes on an I²C bus all have a unique address and only the master can initiate communication. To do so, it issues a start condition to get the attention of the slave devices on the bus. Then the master sends the address of the slave and the wanted data direction, read or write. In the case of a write operation the master continues by sending the data, bit by bit, synchronized to the clock. For a read operation the master only provides the clock while the slave fills in the blanks with its data. The receiver of a byte (the slave or the master) responds with an acknowledge bit so that the transmitter of the byte (the master or the slave) knows that the byte was not lost in cyber space. Sending one byte thus requires nine clock pulses. When all the data has been transmitted, the master issues a stop condition and the bus returns to its idle state, ready for another transaction. Exchanges can be quite elaborate with sequences of reads and writes and even role changes along the way. Several types of transaction errors are defined together with the techniques to recover from these errors and in the end correctly managing an I²C bus can be quite complicated.

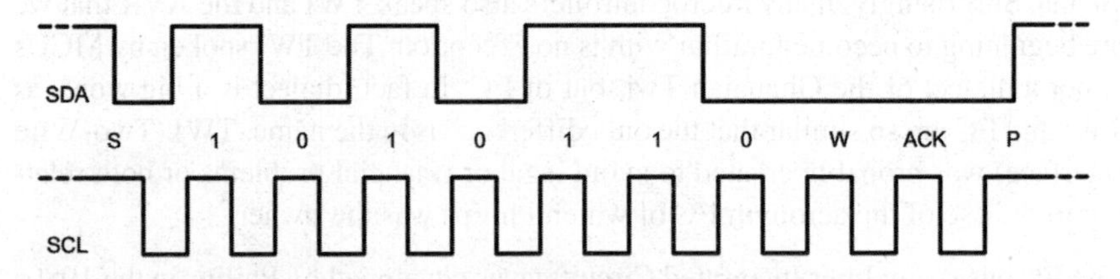

Figure 8-6 - Timing diagram of the I²C bus showing all the common states. 'S' is the start condition and the 7-bit slave address is 0x56. Bit 8 ('W') is low, so this is a write operation and since bit 9 bit (ACK) is low, the slave has acknowledged the command. Finally, the master has terminated the transaction prematurely by issuing a stop condition ('P').

8.4.1 I²C, TWI and Arduino

In Arduino, the I²C bus (or TWI) is managed by the library `Wire` which therefore is not part of the API. The library is based on the `Stream` object and it exposes an object also named `Wire`. If you look at the source code of the library[1], you will see that the object used internally is called `TwoWire`, which is more explicit, but for some reason the designer(s) of the library chose to name the public object `Wire`. This designation can be confusing because there exists another quite popular communication protocol named 1-Wire that uses only one wire (see also Section 10.11.1, page 319).

1. In the folder <arduino>\hardware\arduino\cores\arduino.

```
#include <Wire.h>

void setup()
{
  Wire.begin();
}

void loop()
{
  Wire.beginTransmission(4);    // communicate with device No. 4
  Wire.write("12345");          // send five bytes
  Wire.endTransmission();       // stop sending

  Wire.requestFrom(2,6);        // request 6 bytes from slave No. 2
  while (Wire.available()!=0)
  {
    char c = Wire.read();       // receive a byte
    Serial.print(c);            // transmit the byte on the serial port
  }

  delay(500);
}
```

Because `Wire` is a library, you must import it in your sketch first before you can use the I²C port. Use the option `Import Library...` from the `Sketch` menu to do this. The line `#include <Wire.h>` will then be inserted at the top of your sketch (see also Section 8.1.1, page 160). Call the library's function `begin` in the function `setup`. As explained above, the I²C bus is based on the master-slave model. In many cases the MCU will be the master and specifying its address is pointless; the addresses are only needed when a master wants to talk to a slave. Once the bus is initialized you can begin to send data or commands to a slave (write a slave). Asking a slave to send you data is also possible (read the slave). However, the procedures to use are not identical. Writing a slave device begins with a call to `beginTransmission`, followed by sending the data using the function `write`. The transaction is completed by a call to `endTransmisison`. Reading a slave device starts with a call to `requestFrom`. From then on the procedure is similar to receiving data on the serial port, that is to say, watch (poll) the port using the function `available` and wait for the requested data to arrive. When the data has arrived, you can read it with the function `read`.

The functions `beginTransmission` and `requestFrom` take the address of the slave as an argument; the function `requestFrom` also needs to know how many bytes you would like to receive.

8.4.2 Atmospheric Pressure Sensor

Now is a good moment to put the I²C bus and the `Wire` library into practice. For this I decided to use a pressure sensor HP03 from the manufacturer Hope RF. This sensor exists in several models with different packages and accuracies, and mine is an 'S'-type. If the sensor is left out in the open, it measures the atmospheric pressure, but it can also be used in other pressure measurement applications. Its range is 300 to 1100 millibars (or hPa or mbar). The component must be powered from 3.3 V. Although the Arduino is able to provide this voltage, the board's MCU itself operates at 5 V and generates 5 V signals. This poses a small problem, because the sensor's inputs do not support voltages above 3.6 V, forcing us to use level adapters. This is not really a problem when the I/O port is unidirectional but, in the case of an I²C bus, the SDA line is a bidirectional connection. Furthermore, if more than one device can be the master, the SCL line will be bidirectional too. Fortunately, there exists an elegant solution to this problem, and it requires only one MOSFET per bidirectional signal.

The pressure sensor has two additional unidirectional inputs XCLR and MCLK that we must drive too. A network of a resistor and a Zener diode can limit the voltage levels for these inputs.

On the Uno board the pins of the I²C bus are shared with the two analog inputs A4 (SDA) and A5 (SCL); on the Mega the I²C bus is connected to contacts 1 and 2 of the Communication connector. The XCLR and MCLK signals can be connected to any available output on the MCU.

The datasheet of the pressure sensor is not very explicit. In fact, if you read it carefully you will discover that the component integrates not one but two I²C devices: the sensor (two actually) and an EEPROM that contains calibration values. The EEPROM is read-only. Each component has its own I²C address: 0x77 for the sensor and 0x50 for the EEPROM. The datasheet mentions the addresses 0xee and 0xa1, but these include bit 0, the read/write bit. To obtain the real I²C addresses we must shift all the bits of the two addresses one position to the right, giving 0x77 and 0x50. The read/write bit is added by the `Wire` library. The sensor itself combines two sensors, a temperature sensor and a pressure sensor. The first is used to correct the values obtained with the second. The manufacturer's documentation describes the procedure to follow to apply this correction.

The XCLR and MCLK signals are required only when reading the sensor. In this case, XCLR must be kept high and MCLK must receive a clock signal of about 32 kHz. XCLR must be kept low in order to read the calibration parameters stored in the EEPROM, while MCLK does not seem to have a function in this case. As noted above, XCLR and MCLK can be connected to any digital output of the

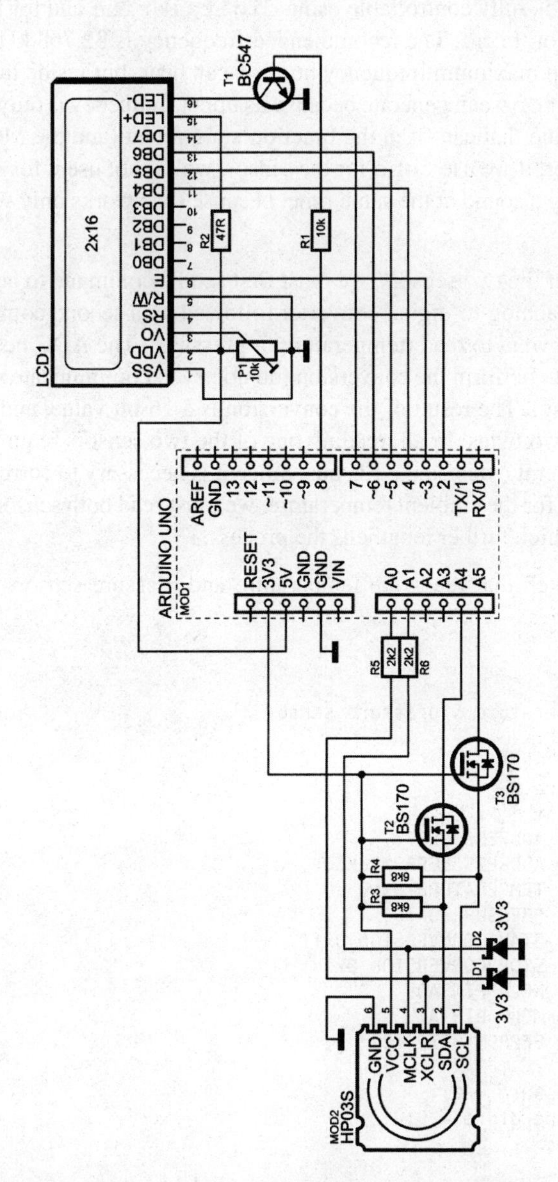

Figure 8-7 - The barometer circuit based on a HP03 sensor from Hope RF. Most of the components to the left of the Arduino board are needed to adapt the levels from 5 V to 3.3 V (and vice versa).

MCU. XCLR is fully controllable using digitalWrite and for MCLK we can use the function tone. The recommended frequency is 32.768 kHz, a frequency well above the maximum frequency humans can hear, but inside the range of frequencies that tone can generate because its upper frequency is only limited by the maximum value that can fit in the function's argument (and the MCU's capabilities). However, if we use tone for the sensor, we cannot use it for something else like producing a sound at the same time, because tone works only with one output at a time.

To read one of the two sensors, we must first send a command to activate the sensor's built-in analog-to-digital converter followed by a second command to select the sensor we want to read (temperature or pressure). The ADC needs some forty milliseconds to perform the conversion (do not try to communicate with the sensor when it is busy). The result of the conversion is a 16-bit value, and so it is necessary to read two bytes. In all, reading one of the two sensors requires three transactions of several bytes on the I²C bus. Since it is necessary to correct the pressure measurement for the ambient temperature, we must read both sensors to obtain the best result, which further lengthens the process.

Here is a sketch that reads the temperature and pressure sensors of the HP03S device:

```
/*
 * HP03 temperature & pressure sensor
 */

#include <Wire.h>

#define HP03_ADDRESS 0x77
#define HP03_ADDRESS_EEPROM 0x50
#define HP03_TEMPERATURE 0xe8
#define HP03_PRESSURE 0xf0
#define HP03_START_CONVERSION 0xff
#define HP03_STOP_CONVERSION 0xfd
#define HP03_XCLR_PIN A0
#define HP03_MCLK_PIN A1
#define HP03_EEPROM_SIZE 18

uint32_t pressure = 0;
uint16_t temperature = 0;
int32_t C[8];
uint8_t AA;
uint8_t BB;
uint8_t CC;
uint8_t DD;
```

```
void HP03_read_eeprom(void)
{
  uint8_t buffer[HP03_EEPROM_SIZE];
  uint8_t i = 0;
  uint8_t j = 0;

  Wire.beginTransmission(HP03_ADDRESS_EEPROM);
  Wire.write(16);
  // Restart without generating a start condition.
  Wire.endTransmission(false);
  Wire.requestFrom(HP03_ADDRESS_EEPROM,HP03_EEPROM_SIZE,true);
  while (Wire.available()!=0 && i<HP03_EEPROM_SIZE)
  {
    buffer[i++] = Wire.read();
  }

  j = 0;
  for (i=1; i<8; i++)
  {
    C[i] = buffer[j++];
    C[i] <<= 8;
    C[i] |= buffer[j++];
  }
  AA = buffer[j++];
  BB = buffer[j++];
  CC = buffer[j++];
  DD = buffer[j++];
}

uint16_t HP03_read_value(uint8_t sensor)
{
  uint16_t value = 0;

  // launch a conversion.
  Wire.beginTransmission(HP03_ADDRESS);
  Wire.write(HP03_START_CONVERSION);
  Wire.write(sensor);
  Wire.endTransmission(true);

  delay(40);

  // Get the result.
  Wire.beginTransmission(HP03_ADDRESS);
  Wire.write(HP03_STOP_CONVERSION);
  // Restart without generating a start condition.
  Wire.endTransmission(false);
  Wire.requestFrom(HP03_ADDRESS,2,true);
  if (Wire.available()!=0) value = Wire.read() << 8;
  if (Wire.available()!=0) value |= Wire.read();
  return value;
}
```

```
void HP03_read(void)
{
  digitalWrite(HP03_XCLR_PIN,HIGH);
  tone(HP03_MCLK_PIN,32768);

  delay(2);

  pressure = HP03_read_value(HP03_PRESSURE);
  pressure += HP03_read_value(HP03_PRESSURE);
  pressure += HP03_read_value(HP03_PRESSURE);
  pressure += HP03_read_value(HP03_PRESSURE);
  pressure /= 4;
  temperature = HP03_read_value(HP03_TEMPERATURE);

  noTone(HP03_MCLK_PIN);
  digitalWrite(HP03_XCLR_PIN,LOW);
}

void HP03_calculate(int32_t& t, int32_t& p)
{
  int32_t offset;
  int32_t sensitivity;
  t = temperature;
  t -= C[5];
  if (t>=0)
  {
    t -= (((t*t)>>14)*AA)>>CC;
  }
  else
  {
    t -= (((t*t)>>14)*BB)>>CC;
  }
  offset = (C[2] + (((C[4]-1024)*t)>>14)) * 4;
  sensitivity = C[1]+((C[3]*t)>>10);
  p = pressure;
  p = ((sensitivity*(p-7168))>>14) – offset;
  p = ((p*10)>>5) + C[7];
  t = 250 + ((t*C[6])>>16) – (t>>DD);
}

void setup(void)
{
  pinMode(HP03_XCLR_PIN,OUTPUT);
  digitalWrite(HP03_XCLR_PIN,LOW);
  pinMode(HP03_MCLK_PIN,OUTPUT);
  digitalWrite(HP03_MCLK_PIN,LOW);
  Serial.begin(115200);
  Wire.begin();
  HP03_read_eeprom();
}
```

```
void loop(void)
{
  int32_t t;
  int32_t p;

  HP03_read();
  Serial.print("T=");
  Serial.print(temperature);
  Serial.print(", P=");
  Serial.print(pressure);
  HP03_calculate(t,p);
  Serial.print(", T=");
  Serial.print((float)t/10.0);
  Serial.print(", P=");
  Serial.println((float)p/10.0);
  delay(1000);
}
```

This sketch perfectly follows the instructions of the manufacturer of the sensor. I will not explain every line; by now you should know enough about Arduino to understand the parts that I will skip. We begin reading, as usually, in the function `loop`. Calling the `HP03_read` function reads the two sensors to obtain the raw temperature and pressure values. `HP03_calculate` transforms these values in an ambient temperature in degrees Celsius and a pressure in hectopascals (hPa, basically the same as millibars) corrected for the ambient temperature. The raw and processed values are sent to the serial port. Because the calculated values are ten times too large, they are divided by ten before sending them.

Correcting the pressure is possible only after having read a number of calibration parameters stored by the manufacturer in the EEPROM of the sensor. They are read in the function `setup` with a call to `HP03_read_eeprom` but not before having activated the I²C bus with `Wire.begin`.

`HP03_read` activates the XCLR line and the clock signal on MCLK. Then the pressure is read four times, and averaged. The manufacturer does not explain why this is necessary; apparently the signal can be sensitive to noise. After reading the temperature the XCLR and MCLK signals are disabled.

The values obtained with `HP03_read` are processed by the function `HP03_calculate`. The seemingly incomprehensible calculation carried out here follows the datasheet almost verbatim, I even kept the names of the parameters C1 to C7 (but as table entries; note that `C[0]` is not used) and AA, BB, CC and DD. It is a kind of calculation in fixed-point arithmetic where I used shift operations (>>) to replace the divisions by powers of two. Note how I put parentheses around almost everything; this is highly recommended for the shift operators << and >>.

Be careful when you mix signed and unsigned data types like I did here. A signed value uses the most significant bit to store the sign; if this bit is equal to one, the value is negative. For example, for a variable of type `int16_t` a value of 0x8000 corresponds to –32,768 while the value 0x7fff (i.e. 0x8000 – 1) equals to +32,767. However, in an unsigned value the most significant bit is like any other bit. For a variable of type `uint16_t` the value 0x8000 equals +32,768. When an unsigned data type is used in a calculation that also uses signed data types, it sometimes happens that the unsigned data type is transformed into a signed data type due to the way the calculation was written down. If this happens when its value's most significant bit is set to one (i.e. when the value is equal to or greater than +32,768) it will be considered a negative value, which is obviously incorrect (the inverse may happen too, a negative value being interpreted as a large positive value). This is the reason why I cut the calculation `t = temperature - C[5]` in two. Doing things this way I made sure that the signs are always preserved. I did the same for the pressure p. It is for similar reasons that the parameters `C[x]` are stored in an array of type `int32_t` instead of in an array of type `uint16_t`. It takes twice as much memory, but avoids unsigned data being misinterpreted as signed. Such problems can be circumvented by putting typecasts everywhere, a technique which unfortunately does not improve the readability of the program.

The function `HP03_read_value` reads one of the two sensors, and it is here that the I²C bus is really used. As noted above, reading a sensor requires two steps, starting a conversion and reading the result. A conversion takes about 40 ms. A conversion is started by writing two bytes to the device, a control byte and a byte to select the temperature or pressure sensor. The procedure to read the result of the conversion is a bit unusual. We must first send the stop command to the converter, i.e. write the device, and then switch to read mode without issuing a stop condition. The function `Wire.endTransmission` with `false` as argument allows us to do this. After this call we can continue the communication in read mode to obtain the two bytes that constitute the raw measured value.

The function `HP03_read_eeprom` uses the same technique to read the eighteen bytes of the eleven calibration parameters. These bytes are placed in a buffer before being collected and copied into the array `C[]`. Note that `C[0]` is not used at all, and so the table is slightly too big. The parameters `AA`, `BB`, `CC` and `DD` occupy one byte each.

Since the sketch transmits the measured value over the serial port, we can use Arduino's serial port monitor to view the results. At the end of this chapter we will add an alphanumeric LCD to the barometer (and also a humidity sensor), but nothing prevents you to do this now and it is a good exercise. For my part, I will continue with the SPI port to which I plan to connect a color graphic display.

8.5 Three- and Four-Wire Connections

The Serial Peripheral Interface (SPI, also called Spy) is a link comprising four wires: inbound data (MISO, DI or SDI), outbound data (MOSI, DO or SDO), clock (CLK or SCK) and select (SS or CS). It seems that the acronym SPI was legally deposited by Motorola (today Freescale), but the protocol never got standardized. SPI is a *de facto* standard, which explains why so many variations exist today. The differences mainly concern in the packet length, the polarity of the signals and the phase between them. Microwire or μWire by National Semiconductor (now Texas Instruments) is very similar to SPI. A variant named QSPI (or Microwire/Plus) that adds receive buffers exists also. When the MISO (Master In Slave Out) and MOSI (Master Out Slave In) signals are combined to create a single signal SISO (Slave In Slave Out), we end up with a three-wire SPI bus.

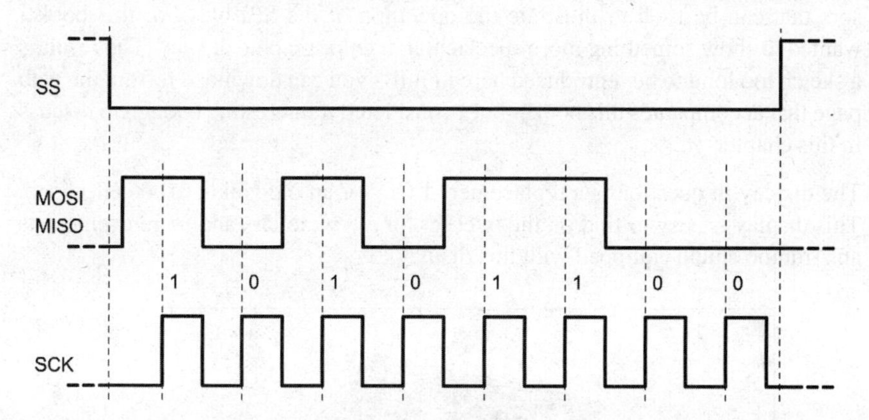

Figure 8-8 - Typical timing diagram for an SPI bus. Here the MISO or MOSI line is sampled on the rising edges of the SCK signal.

The success of SPI is due to the high data speeds possible and the small number of pins required. In fact, SPI is a good alternative to the parallel port and you can find this kind of interface on components that handle large amounts of data. SPI is also very easy to implement purely in software because the precision of the clock is often of little or no importance, making this a very flexible communication bus.

On first sight I²C and SPI look very similar – both are synchronous links based on the master-slave paradigm – but there are many important differences. SPI adds a selection signal (meaning an extra wire) to enable the receiver, where I²C uses addresses. I²C is a link that cannot be used simultaneously in both directions (it is half-duplex), whereas SPI allows full-duplex communication using the MISO and

MOSI signals. In an SPI connection the synchronization signal or clock can be provided by the master or by the slave and not only by the master as is the case for I^2C. The latest version of the I^2C standard allows a maximum speed of 5 MHz; the SPI bus can handle 50 MHz or more without any problems. Contrary to I^2C, SPI does not impose a specific communication protocol, each manufacturer implements whatever suits him best, and it is not uncommon that a program is forced to continually reconfigure the SPI bus in a system which incorporates devices from several manufacturers. On the hardware side, I^2C and SPI are similar enough to allow a combination of the two busses in a single peripheral module and you can end up with a simplified I^2C protocol over an SPI connection.

8.5.1 Improved Driver for Graphic Display

There are plenty of electronic components like memories, data converters or sensors that can be used to illustrate the operation of the SPI bus. For this book I wanted to show something more spectacular: a color graphic display. The result is a sketch too long to be reproduced here in full – you can download it from the web page that accompanies this book – but I considered it interesting enough to discuss in this chapter.

The display in question is a replacement LCD for an old Nokia 6100 cell phone. This display is easy to find on the internet for a few dollars and there even exists an Arduino shield equipped with this display.

Figure 8-9 - A shield sporting a mobile phone color graphic display.

The software driver for this display is also available on the internet, including an Arduino version. So if all the work to use this display has already been done and published, why talk about it here? Well, first of all because this example allows me to introduce the SPI peripheral modules. More important however is the fact that I can use it to show you a modification that not only highlights the flexibility of the SPI bus at the timing level, but that also triples the execution speed of the driver. This mod does not allow you to watch streaming video on an Arduino board, but it will be possible to make a simple digital oscilloscope or a logic analyser, which is not bad, right?

This display, which comes in two versions, Epson and Philips, uses nine-bit words to communicate. This ninth bit is used to differentiate the data (mainly the pixels) from the commands (mainly the initialization). However, the SPI module of the AVR can only handle eight-bit words; it does not know how to deal with the ninth bit[1]. To work around this problem, the author of the driver has written a software SPI device using a technique called bit banging. Bit banging means that the signals are produced by manipulating pin levels “manually”. Here is the function to send data to the display (the driver has a second almost identical function to send a command):

```
void Nokia6100_send_data(uint8_t data)
{
  uint8_t i;

  digitalWrite(CS,LOW);

  digitalWrite(SISO,HIGH);
  digitalWrite(SCK,LOW);
  digitalWrite(SCK,HIGH);

  for (i=0; i<8; i++)
  {
    if ((data&0x80)==0x80)
    {
      digitalWrite(SISO,HIGH);
    }
    else
    {
      digitalWrite(SISO,LOW);
    }
```

1. The AVR’s USART module features an SPI mode for sending 16-bit words and it can handle 9-bit data in asynchronous mode, but it cannot deal with 9-bit data in synchronous mode.

```
    digitalWrite(SCK,LOW);
    digitalWrite(SCK,HIGH);

    data <<= 1;
  }

  digitalWrite(CS,HIGH);
}
```

Note that for clarity I used functions of the Arduino API to avoid introducing too many new things here. This version is much, much slower than the real thing.

The display uses the same pin to send and receive data, so it is a three-wire SPI bus (`CS`, `SCK` and `SISO`). Words are sent with their most significant bit (MSB) first and the clock `SCK` is active low. The function starts by activating the display using the `CS` signal. Then it transmits the ninth bit, a '1', to indicate that it is sending data and not a command. To do this, the `SISO` line is high and a pulse is generated on the `SCK` line (active low). The algorithm assumes that `SCK` was high when the function was entered. The next eight bits are processed in a loop. If the MSB of the data is '1' (`(data&0x80)==0x80`), `SISO` is set to `HIGH`; if the bit is '0', `SISO` is set to `LOW`. The bit is sent by briefly activating the `SCK` signal (generating a strobe). Then the remaining data bits are shifted one position to the left for the next pass through the loop. The most significant bit that was just transmitted drops out of `data`, the new LSB is a '0'.

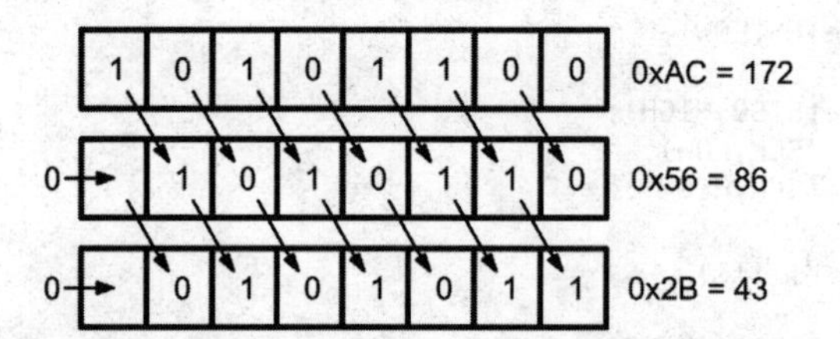

Figure 8-10 - Shifting bits or a shift register. At each shift we lose a bit, but we gain a zero. The value is divided by two.

This function works perfectly fine, but is way too slow. To get something usable, we must replace `digitalWrite` by a function that is as fast as possible and it is best to write it in assembly language as did the author. The MCU's SPI peripheral is not used at all, which is a bit of a shame. That said, a peripheral written in software can use any pin, making it easier to connect the display.

Now we come to the modification that accelerates this function dramatically. It is a simple yet effective mod and it is only possible because of the SPI bus' flexibility with respect to the frequency of the SCK signal that synchronizes the communication. The trick is to use the MCU's SPI peripheral in place of the loop that bit-bangs the eight data bits. The ninth bit is still sent manually.

```
void Nokia6100_send_data_fast(uint8_t data)
{
  digitalWrite(CS,LOW);

  digitalWrite(SISO,HIGH);
  digitalWrite(SCK,LOW);
  digitalWrite(SCK,HIGH);

  SPI.begin();
  SPI.transfer(data);
  SPI.end();

  digitalWrite(CS,HIGH);
}
```

As before I used only Arduino API functions to which I added the SPI library. The function SPI.transfer replaces the for loop of the previous function that now uses the on-chip SPI peripheral. SPI.begin and SPI.end ensure that the SPI peripheral is active only when we need it, otherwise it will not be possible to send the ninth bit manually. This second version is not only faster than the first; it also consumes less program memory. Icing on the cake, the AVR's SPI module features a "double SPI speed" option which allows an 8 MHz clock on an Arduino board. With this option enabled, we get a pretty good communication speed.

The real function is obviously more efficient, especially when it comes to calling SPI.begin:

```
#define cs_on() cbi(CS_PORT,CS_BIT)
#define cs_off() sbi(CS_PORT,CS_BIT)
#define siso_lo() cbi(SISO_PORT,SISO_BIT)
#define siso_hi() sbi(SISO_PORT,SISO_BIT)
#define sck_lo() cbi(SCK_PORT,SCK_BIT)
#define sck_hi() sbi(SCK_PORT,SCK_BIT)

void Nokia6100_send_data_really_fast(uint8_t data)
{
  cs_on();

  siso_hi();
  sck_lo();
  sck_hi();
```

```
  SPCR |= _BV(SPE);                 // SPI.begin
  SPDR = data;
  while (!(SPSR & _BV(SPIF)));
  SPCR &= ~_BV(SPE);                // SPI.end

  cs_off();
}
```

In this code fragment I have defined some macros to improve the readability of the function. After compilation, these macros are each replaced by a single assembly language instruction `sbi` (set bit) or `cbi` (clear bit).

The first four lines and the last one are one-on-one translations of the previous `digitalWrite` statements. The `SPI.begin` command is replaced by the activation of the SPI module by setting the SPE (SPI enable) bit to '1' in the control register SPCR (SPi Control Register). The `SPI.end` replacement right after the while loop cancels this operation by clearing the bit to '0'. The contents of the variable `data` are sent as soon as they are copied into the data register SPDR (SPi Data Register). The **`while`** loop waits for the end of the transmission by monitoring the SPIF bit (SPi Interrupt Flag) of the status register SPSR (SPi Status Register). As soon as this bit is set to '1', the loop will be terminated. The SPIF bit indicates that an interrupt request was generated if interrupts were enabled. Interrupts are discussed in detail in Chapter 10.

8.5.2 Humidity Sensor

The previous example already showed how the SPI bus is malleable, but it is possible to do better (or worse, if you prefer). Indeed, after many hours of trawling the internet I managed to find a component that perfectly illustrates how the SPI bus is not standard at all. The device in question is the temperature and humidity sensor SHT11 from Sensirion. The component provides an interface electrically compatible with the I²C bus, but communicating using a custom protocol inspired by both I²C and SPI. The protocol uses two signals data (DATA) and clock (SCK). The DATA signal is bidirectional and the master is supposed to provide a clock signal (like I²C) with an undefined frequency (like SPI) that must not exceed 5 MHz (like I²C, and if the component is powered from 5 V). There is no slave select signal (unlike SPI), and there is no slave address either (unlike I²C). In fact, a special bit sequence is used to get the attention of the device. Like the SPI bus, selecting between reading and writing is done using commands sent by the master, and not with a reserved bit as on the I²C bus. However, a reserved bit is used to confirm the reception of a command or a data byte, as on the I²C bus. The data sent by the sensor is complemented by a byte that contains a cyclic redundancy check or CRC, which allows to check whether the data has been received correctly. This byte is like the checksum in an NMEA 0183 sentence, but calculated differently.

The manufacturer of the component provides a programming example in C showing how to communicate with the sensor and I have based my sketch on that. Unfortunately, the example does not show how to calculate the CRC, but for you, dear reader, I made an effort to add it to the sketch that you can download.

On the hardware side, the SHT11 sensor can be connected to an I²C bus without disturbing other bus users and without being bothered by them. For this to work, the MCU must be able to disconnect the hardware I²C bus interface to make room for a software interface. The Arduino MCU can do this, so I attached the sensor to the I²C bus which was already connected to the pressure and temperature sensor HP03 (see Section 8.4.2, page 190). The recommended SHT11 supply voltage is 3.3 V (which offers the best accuracy), but it can also run from 5 V and in this case the communication speed may be higher because the clock may go up to 5 MHz (against 1 MHz for a supply voltage below 4.5 V). Although my system disposes of a 3.3V I²C bus thanks to the level adapters, I decided to connect the SHT11 directly to the MCU. This way it works without level adapters, which greatly simplifies the circuit in case of a hygrometer-only setup.

Below is the sketch to read both the temperature and humidity sensor integrated in the device. I simplified it a bit to show only the essential parts, otherwise it would take too much space in this book. The sketch in all its splendor is available for download on the web page that accompanies this book. Note that this circuit does not use the HP03 pressure sensor, but it is not hindered by the presence of it either. One final note before we start reading the sketch: the humidity sensor is available in several models (SHT10, SHT11 and SHT15 that all have different accuracies), which is why I prefixed all the functions with SHT1x instead of with SHT11.

```
#define SHT1X_CMD_READ_TEMPERATURE 0x03
#define SHT1X_CMD_READ_REL_HUMIDITY 0x05

const float C1 = -2.0468;           // for 12 Bit RH
const float C2 = +0.0367;           // for 12 Bit RH
const float C3 = -0.0000015955;     // for 12 Bit RH
const float T1 = +0.01;             // for 12 Bit RH
const float T2 = +0.00008;          // for 12 Bit RH

int data = A4;
int sck = A5;

void strobe(void)
{
```

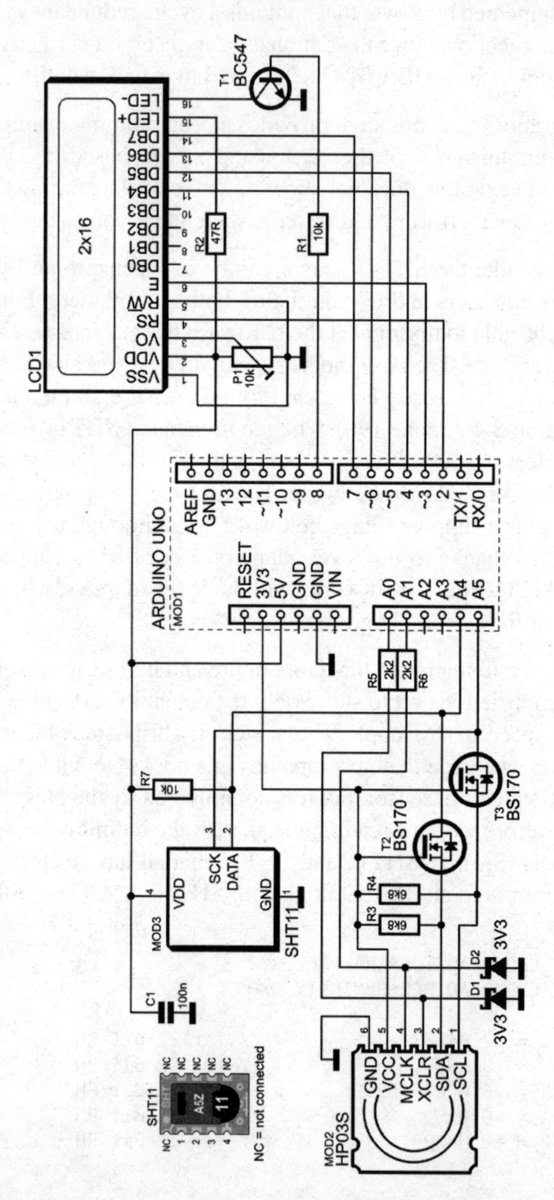

Figure 8-11 - Schematic of the hygrometer based on an SHT11sensor from Sensirion. The sensor is compatible with the I²C bus, even though it does not speak I²C.

```
  digitalWrite(sck,HIGH);
  delay(1);
  digitalWrite(sck,LOW);
}

void SHT1x_start_sequence(void)
{
  pinMode(data,OUTPUT);
  pinMode(sck,OUTPUT);
  digitalWrite(data,HIGH);
  digitalWrite(sck,LOW);
  digitalWrite(sck,HIGH);
  digitalWrite(data,LOW);
  digitalWrite(sck,LOW);
  digitalWrite(sck,HIGH);
  digitalWrite(data,HIGH);
  digitalWrite(sck,LOW);
}

void SHT1x_connection_reset(void)
{
  int i;
  pinMode(data,OUTPUT);
  pinMode(sck,OUTPUT);
  digitalWrite(data,HIGH);
  digitalWrite(sck,LOW);
  for (i=0; i<9; i++)
  {
    strobe();
  }
  SHT1x_start_sequence();
}

boolean SHT1x_send_byte(uint8_t value)
{
  uint8_t i;
  boolean error = true;

  pinMode(data,OUTPUT);
  pinMode(sck,OUTPUT);

  for (i=0x80; i>0; i>>=1)
  {
    if ((value&i)!=0) digitalWrite(data,HIGH);
    else digitalWrite(data,LOW);
    strobe();
  }

  pinMode(data,INPUT_PULLUP);
  digitalWrite(sck,HIGH);
  if (digitalRead(data)==0) error = false;
```

```
  digitalWrite(sck,LOW);
  return error;
}

uint8_t SHT1x_receive_byte(uint8_t ack)
{
  int i;
  uint8_t value = 0;

  pinMode(data,INPUT_PULLUP);

  for (i=0x80; i>0; i>>=1)
  {
    digitalWrite(sck,HIGH);
    if (digitalRead(data)!=0) value |= i;
    digitalWrite(sck,LOW);
  }

  pinMode(data,OUTPUT);
  digitalWrite(data,ack);
  strobe();
  digitalWrite(data,HIGH);

  return value;
}

boolean SHT1x_wait_for_completion(void)
{
  pinMode(data,INPUT_PULLUP);
  while (digitalRead(data)!=0);
  return true;
}

boolean SHT1x_read(int& result, uint8_t& crc, uint8_t command)
{
  boolean error;
  uint8_t temp;
  SHT1x_start_sequence();
  error = SHT1x_send_byte(command);
  SHT1x_wait_for_completion();
  temp = SHT1x_receive_byte(LOW);
  result = temp << 8;
  temp = SHT1x_receive_byte(LOW);
  result += temp;
  crc = SHT1x_receive_byte(HIGH);
  return error;
}

void SHT1x_calculate(float& humidity, float& temperature)
{
  temperature = temperature*0.01 - 40.1;
```

```
  humidity = C3*humidity*humidity + C2*humidity + C1;
  humidity = (temperature-25)*(T1+T2*humidity) + humidity;
  humidity = constrain(humidity,0.1,100);
}

void setup(void)
{
  Serial.begin(115200);
  SHT1x_connection_reset();
}

void loop(void)
{
  int t;
  int h;
  float temperature;
  float humidity;
  uint8_t crc;
  boolean error;

  error = SHT1x_read(t,crc,SHT1X_CMD_READ_TEMPERATURE);
  error |= SHT1x_read(h,crc,SHT1X_CMD_READ_REL_HUMIDITY);
  if (error==false)
  {
    temperature = t;
    humidity = h;
    SHT1x_calculate(humidity,temperature);
    Serial.print("T=");
    Serial.print(temperature);
    Serial.print(", RH=");
    Serial.println(humidity);
  }

  delay(1000);
}
```

This sketch is similar to the one that read the pressure sensor HP03 with a function to read the temperature and the humidity and a function to correct the humidity for the ambient temperature (not forgetting, to correct the temperature for the supply voltage). It also sends the resulting data to the serial port. Since we are dealing with a software device, the sketch also contains functions for sending and receiving bits on the home-made SPI bus.

Again, we begin reading in the function `loop`. The raw temperature and humidity values are read first using the `SHT1x_read` function. Each read operation may produce an error, so data processing is performed only if both reading operations were executed without problems. The processing consists of a calculation provided by the manufacturer, I only translated it into C. The raw values are stored in integers; the calculation uses floating-point numbers. The way I pass arguments to the function `SHT1x_calculate` forced me to copy the raw values in variables of type **`float`**.

The `SHT1x_read` function begins with a call to `SHT1x_start_sequence` that generates pulse sequences on the DATA and SCK lines meant to wake up the device. I suggest you draw the timing diagram of the two signals to better understand `SHT1x_start_sequence`; you can compare your drawing with the one in the datasheet.

The start sequence is followed by the command to read the temperature or humidity sensor. Supposing that the device has understood the command, the sketch will now wait for the completion of the command. This event is flagged to the MCU by the SHT1x device pulling its DATA line down. As soon as the MCU detects this, it can read the two bytes that make up the 16-bit measurement. It is not really necessary, but it would be correct to also read the third byte that contains the checksum (CRC). This last read operation completes the transaction. In this sketch the CRC is not used, but the sketch available for download shows how to verify the checksum.

As already mentioned above, the calculation performed by `SHT1x_calculate` is explained in the datasheet of the component, and the values of the constants `C1`, `C2`, `C3`, `T1` and `T2` can be found there too.

Both functions `SHT1x_send_byte` and `SHT1x_receive_byte` are low-level, meaning that they are involved in the transmission and reception of the bits on the DATA and SCK lines. The DATA line is an output in `SHT1x_send_byte` and an input with pull-up resistor in `SHT1x_receive_byte`. SCK is an output in both cases. The two functions use a loop to process all eight bits[1]. After sending or receiving eight bits, the DATA line changes direction. It becomes an input with a pull-up resistor in `SHT1x_send_byte` and an output in `SHT1x_receive_byte`. This is required so that it can send or receive the acknowledge bit. The function `strobe` produces the pulses on the clock line SCK.

We conclude our code review with the function `SHT1x_connection_reset` that sends a minimum of nine clock pulses with the DATA line set to '1' to clear the receive buffer of the SHT1x. A new transaction between the component and the MCU may now be initiated.

1. If the sensor would have been connected to the MCU's SPI bus, this part could have been handled by the SPI module.

8.6 All Together

Now that we have connected two sensors and a display to the Arduino board, it would be interesting to write a sketch that brings everything together in a simple weather station application. To avoid ending up with a sketch of several pages long, I created Arduino libraries for the HP03 and SHT1x devices. Thanks to these libraries, the sketch is concise and we can focus on essentials, that is to say, reading and displaying the temperature, the atmospheric pressure and the humidity. To create a library you should, basically, create two files with the same name but with different extensions and put them in a subdirectory of the same name inside the libraries subdirectory of the Arduino installation or in the libraries subdirectory of the Sketchbook location (see Section 4.4.1, page 47). For example, to make a library for the HP03 component, the files `HP03.cpp` and `HP03.h` must be created and put in the subdirectory <arduino>\libraries\HP03. Then you have to copy all the functions that should be part of the library in the `HP03.cpp` file and all the function prototypes in the file `HP03.h`. A function prototype consists of the name and type of the function and of its arguments, all followed by a semicolon. For example, the prototype of the function `HP03_calculate` is:

```
void HP03_calculate(int32_t& t, int32_t& p);
```

This is not rocket science. In general, the prototype of a function is the first line of the function terminated with a semicolon.

Creating your own libraries is actually a bit more involved than I just made you believe, and requires a deep knowledge of C and C++. If you're interested, I advise studying my libraries as well as those supplied with Arduino.

Here is the sketch for the weather station that uses the `SHT1x` and `HP03` libraries:

```
/*
 * weather3
 * display temperature, pressure & humidity
 */

#include <LiquidCrystal.h>
LiquidCrystal lcd(2,3,4,5,6,7);

#include <SHT1x.h>
SHT1x sht1x;

#include <Wire.h>
#include <HP03.h>
HP03 hp03;
```

```
#define __CELSIUS__
// custom character.
uint8_t degree[8] =
{
#ifdef __CELSIUS__
  0b11100,
  0b10100,
  0b11100,
  0b00000,
  0b00011,
  0b00100,
  0b00100,
  0b00011
#else
  0b11100,
  0b10100,
  0b11100,
  0b00000,
  0b00111,
  0b00100,
  0b00110,
  0b00100
#endif /* __CELSIUS__ */
};

void setup(void)
{
  lcd.createChar(0,degree);
  lcd.begin(16,2);
  lcd.print("Arduino Weather");
  delay(750);
  lcd.clear();
  lcd.print(" T   RH  mbar");
}

void loop(void)
{
  int t = 0;
  int p = 0;
  int rh = 0;

  hp03.begin(A1,A0);
  hp03.update();
  t = hp03.get_temperature() + 0.5;
  p = hp03.get_pressure() + 0.5;

  sht1x.begin(SDA,SCL,true);
  sht1x.update();
  t += sht1x.get_temperature() + 0.5;
  t /= 2;
  rh = sht1x.get_humidity() + 0.5;
```

```
#ifndef __CELSIUS__
  t = t*9/5 + 32;                    // Celsius to Fahrenheit.
#endif

  lcd.setCursor(0,1);
  if (t>=0 && t<100) lcd.print(' ');
  lcd.print(t);
  lcd.write((uint8_t)0);
  lcd.print(' ');
  lcd.print(rh);
  lcd.print("% ");
  lcd.print(p);

  delay(1000);
}
```

Because I used libraries, this sketch remained fairly simple. In the function `loop` we first read the pressure sensor HP03. There is no specific reason for starting with this sensor; let's say that we read the sensors in alphabetical order. The function `hp03.begin` initializes the I²C link. This is necessary for each pass through loop, because the lines of the I²C bus are shared with the SHT11 humidity sensor that is "not I²C". After `hp03.begin` we call `hp03.update` which ensures that the temperature and pressure values provided by the sensor are updated. The functions `hp03.get_temperature` and `hp03.get_pressure` help us to obtain these values. We add 0.5 to the values to ensure that rounding will be correctly done when these floating-point values are copied into integer variables (see also Section 7.1.1, page 118).

We repeat the same operations for the humidity sensor SHT11 (or SHT1x). The temperature measured by this sensor is added to the one obtained with the other sensor and the total is divided by two. The resulting temperature is therefore the average of these two temperatures.

I added an option to convert the temperature calculated in degrees Celsius to degrees Fahrenheit. If you prefer displaying the temperature in Fahrenheit, uncomment the line `#define __CELSIUS__` at the top of the sketch.

Once we have collected the required data, we can print it on the second line of the display; the first line will show a sort of legend. The only thing to notice here is the line `lcd.write((uint8_t)0)`. The purpose of this statement is to show a custom-built character: degree-Celsius or degree-Fahrenheit.

If you examine the function `setup` closely, you will see there a call of the function `lcd.createChar`. This function allows you to define up to eight special characters, from 0 to 7, designed by you. Simply create a table with the definition of the character, like I did with the array `degree[]`. The array contains binary

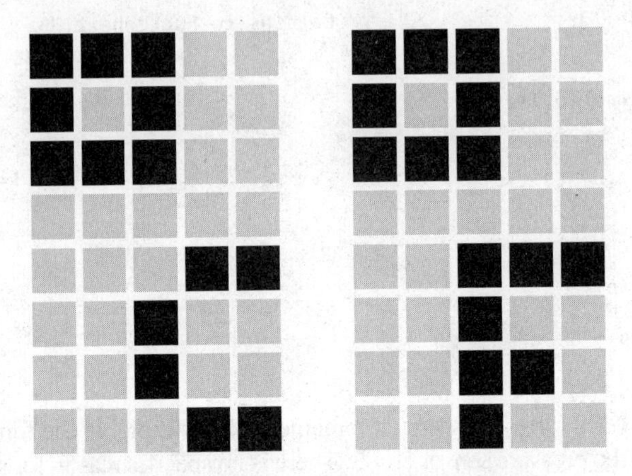

Figure 8-12 - A custom LCD character saves a position on the display.

values – indicated by the prefix `0b` – because it is more descriptive, but if you prefer, you can use hexadecimal values or even plain decimal. A character consists of eight rows of five pixels; a '1' corresponds to a lit pixel (black), a '0' to a pixel that is off (transparent).

The display legend is displayed at the end of `setup`, T means Temperature, RH means Relative Humidity and mbar stands for Pressure.

Done. You now own a nifty mini weather station that you can show to your friends. But do not expect a lot of oohs and aahs or any applause for that matter. Instead, your friends will probably say something like "Do you have any idea how cheap such a thing is on the internet?" or "So that's is why you have snubbed us for three weeks?" and "If you prefer making useless things instead of hanging out with us, you might as well stay home!", and there you are, free again to spend all your time on programing microcontrollers. To keep you busy I propose the following exercise: add, all by yourself, serial port support to the sketch so you can record the weather data on your computer. This allows you to create beautiful graphs showing the evolution over time of the atmospheric pressure, the humidity and the temperature. Your friends will love it!

8.7 When Arduino Isn't Around

Using the API functions and Arduino libraries is very easy, but how to go about when you do not have these tools? Actually, you already know what to do.
Even if you probably did not recognize them as such when the example was presented, all the techniques have been shown. It is the function `Nokia6100_send_data_really_fast` that showed you the way. No need to look it up, I copied the interesting part of this function here for you. The example concerns the SPI device, but the technique used is valid for most serial communication devices.

```
SPCR |= _BV(SPE);                              // enable the module
SPCR = (SPCR & ~SPI_MODE_MASK) | mode;         // select the module's mode
SPCR = (SPCR & ~SPI_CLOCK_MASK) | rate;        // select the communication speed
…
SPDR = data;                              // send a byte
while (!(SPSR & _BV(SPIF)));              // wait for the transmission to end
```

After activating the device, the communication parameters such as the bit rate, the number of bits per word, and similar things have to be specified. The parameters that have to be defined depend of course on the communication module. An SPI port for example has to know what to use for the clock polarity (active high or low), and how the clock is to be synchronized with the data. A serial port needs to know the number of stop bits, and whether or not it is necessary to calculate a parity bit. An I²C module has to know for example whether it is a master or a slave. In short, each device has its own characteristics that depend on the MCU, and I recommend to always consult the user manual to find out the details and subtleties of its peripherals.

All these modules have in common that in order to transmit data, it must be copied to the data transmit register (SPDR in the example). Then you have to wait for the activation of a status bit (flag) which signals the end of the transmission (the SPIF bit of the SPSR register in the example), or wait for an interrupt request to be generated. Once such an event has been observed, the next data byte or word can be sent.

Receiving data is done almost identically. In this case no need to write to a data register, you just have to wait for the activation of the flag or the interrupt that indicates the arrival of new data. The received data will be available in the data receive register (which, in SPI, is often the same as the data transmit register).

Interrupts will be explained in another chapter.

8.8 Pointers

Here is a subject that has forever confused programming newbies: pointers. In addition to that, pointers form a controversial issue that keeps computer scientists and programmers divided into two groups: those in favor and those against. Those who do not know what we are talking about are in between these two groups. But what exactly is a pointer? Well, a pointer is something that is easy to describe, but difficult to use correctly.

Each variable and each instruction of a program get stored somewhere in a location of the system's memory. This memory location has an address, just like your apartment or your house. In a computer, that address is just a number. Like there's just one street. Often there are actually two streets: Data Memory Street, and Program Memory Street. Microcontrollers generally have a third street: Built-in Devices Street. For example, a variable is number 0x4a76 Data Memory Street while the first instruction of a program is often at 0x0000 Program Memory Street. House numbers are always written in hexadecimal notation. A pointer simply contains the address of the variable or the instruction, like the address you write on an envelope to be put in the mail or the e-mail address of one of your friends.

Pointers are typically used to:

1. avoid copying data. Every time you call a function that takes one or more arguments, they are copied into a special memory called the stack (that will be discussed in more detail in Chapter 10). The function works with the copies of the arguments, so you can modify an argument within a function without affecting the original argument. But when an argument consists of many bytes, copying it first costs time and valuable memory space. If, in addition, the special stack memory lacks space, data can be lost and your program may crash. Using a pointer to the argument instead of the argument itself can make a program more efficient;

2. edit anonymous data. Some functions produce multiple results, such as the function `SHT1x_calculate` described before that processes the temperature and the humidity data. With this type of function you may wish to return a result and an error or status code to signal to the caller that a problem was encountered or, on the contrary, that everything went fine. However, a function can return only one result. Here the use of pointers can provide a solution. The function cannot change the argument itself, since it only has a copy of it, but it can change the data that lives at the address specified by a pointer. So if the caller provides the function with the addresses of the arguments to modify instead of the arguments themselves, the function can change them;

3. create lists. Many applications work with lists where each element in the list may contain a certain amount of data. Often the length of the list, the size of its elements or both are not known in advance. How do you reserve memory for such a list without the risk of reserving too much or, worse, not enough? In this case dynamic memory allocation is used, meaning that the list items are created on the fly, when the program is running, somewhere in memory where there is room. A pointer to each newly created item is saved in the item created previously; sometimes also a pointer to the previous item is stored in the new item. This way the items that may be physically dispersed in memory are logically chained together using pointers and form a coherent list;

4. create objects on the fly. This is like creating a list item, but without a list. Since the location in memory where the object is created is not known in advance, the creation of the object returns a pointer containing the address of the new object;

5. call functions. Function names are also pointers.

Pointers can be used in calculations, and if you do not pay attention to its value, a pointer may stop pointing to where you thought it was pointing, resulting in unexpected and often disastrous program behavior. That's why pointers are considered dangerous; they facilitate accidental data modifications. Some programming languages even prohibit the use of pointers. In C however pointers are used extensively because they can accomplish tasks more easily than in other languages.

The ability of C to change with a typecast the data type of the object referenced by a pointer is sometimes convenient, but extremely risky. For example, using this technique you can turn the address of a variable of type **`char`** (one byte) into an address of a variable of type **`float`** (four bytes in Arduino). If you use this modified pointer to increment the variable of type **`char`**, you will modify three additional bytes as well. Needless to say, poorly managed pointers are responsible for the majority of bugs in programs written in C. A classic instance is the pointer that does not point anywhere, because it is equal to zero (the infamous null pointer) or because you forgot to initialize it, leaving it with a random value.

The designers of C++ have tried to improve things a bit by introducing so-called references. These are similar to pointers, but managed by the compiler instead of by the programmer. Contrary to a pointer, a reference cannot be modified or be used in a calculation. However, it is possible to change an argument that was passed "by reference". Here is an example of a function intended to increment the time once per second. First, in C:

```
void clock_tick(int *p_hours, int *p_minutes, int *p_seconds)
{
  *p_seconds += 1;
  if (*p_seconds>59)
  {
    *p_seconds = 0;
    *p_minutes += 1;
    if (*p_minutes>59)
    {
      *p_minutes = 0;
      *p_hours += 1;
      if (*p_hours>23)
      {
        *p_hours = 0;
      }
    }
  }
}

int hours;
int minutes;
int seconds;
clock_tick(&hours,&minutes,&seconds);
```

The asterisk '*' indicates that we are dealing with pointers and not with "normal" variables. By the way, did you notice how I prefixed the function arguments with "p_"? I took this habit to avoid making mistakes in my programs by confusing pointers and normal variables and I highly recommend you to adopt a similar strategy. How to call the function `clock_tick` is shown at the end of the code fragment. The ampersand character ('&') in front of each argument indicates that the argument is actually the address of the variable and not the variable itself. The asterisk cancels the ampersand, so `*&hours` equals `hours`[1], which is why I used the asterisk everywhere in the function. If I had forgotten an asterisk somewhere, the compiler would not have said anything, because all the instructions are legal with or without asterisk. Do you begin to see the danger? For example, had I forgotten the asterisk in the line `if (*p_minutes>59)`, the condition would have been false if the address contained in `p_minutes` was less than or equal to 59, or always true otherwise. The code in the body of the **if**-statement following it would have been either always executed, or never, which was not at all what the programmer had in mind. Consider another example. Imagine that I had omitted the asterisk in the line `*p_minutes += 1`. Now, instead of incrementing the number contained in `minutes`, I would have incremented the address of `min-`

1. `&*hours` is illegal, because `hours` is not a pointer.
   `*&p_hours` is identical to `p_hours`.

utes with the result that p_minutes would reference the integer that lives in the memory just after minutes. The **if**-statement that follows this line would then be testing the wrong variable with unpredictable effects as a result.

Now I can explain the calculation at the end of the example read_nmea_sentence from the beginning of this chapter and which is at the origin of this section. In C, the name of an array without its square brackets "[ ]" is always a pointer (as it the name of a function without its round brackets). The character string **char** sentence[SENTENCE_MAX] from this function is an array of variables of type **char** and sentence contains the memory address of the first element of the array. Because sentence is a pointer, it can be used in calculations. Sentence+10 corresponds to the address of tenth element of sentence and *(sentence+10) is the tenth element, i.e. sentence[10]. You may have guessed it, '*' is identical to "[0]" and *p_minutes is the same as p_minutes[0].

Here is the same function clock_tick in C++, but now the arguments are passed by reference instead of as pointers:

```
void clock_tick(int &hours, int &minutes, int &seconds)
{
  seconds += 1;
  if (seconds>59)
  {
    seconds = 0;
    minutes += 1;
    if (minutes>59)
    {
      minutes = 0;
      hours += 1;
      if (hours>23)
      {
        hours = 0;
      }
    }
  }
}

int hours;
int minutes;
int seconds;
clock_tick(hours,minutes,seconds);
```

Not only is this function much easier to read without the asterisks and the odd prefixes everywhere, but also the risk of errors is much lower because the only place where you can make real mistakes is in the function definition. And if you had forgotten an ampersand somewhere, the effects would be less dramatic. The function would not have given the desired result, but there would not have been any inexplicable side effects either due to the accidental modification of a portion of the memory.

In this book I have tried to never use pointers unless I could not come up with a better way to do things.

8.9 Did you Know?

The subject of communication is so vast that it was impossible for me to cover it completely. Here is a short list of topics possibly beyond the scope of this book, but not uninteresting either.

- Did you know that the RS-422 and RS-485 standards do not specify signals, data rates or numbers of bits, but only the electrical connection part? When you have a UART you can "do" RS-485 (or RS-422) simply by adding a level converter. Indeed, the notion of voltage-with-respect-to-a-common-rail (usually ground) to specify logic levels is here replaced by the voltage difference between the two conductors of the connection. We talk of a symmetrical or differential connection (relative to common) compared to the asymmetrical connection of for example RS-232. The conductors in RS-485 are called A and B, and the logic value on the line is determined by calculating $A - B$ (differential, remember?). By definition $A = -B$, a bit therefore equals $2A$. When $2A$ is less than zero, the level is a logic low; when $2A$ is larger than zero, the level is a logic high[1]. A full-duplex connection requires four wires, two per communication direction (inbound and outbound).
- Did you know that these standards have been introduced to bypass the distance and speed limitations imposed by standards such as RS-232? If you need to connect a serial GPS receiver over a wired connection to a computer that is 1 km (0.6 mile) away, you can use RS-485 or RS-422 transceivers. This type of connection is used extensively in the industry.
- Did you know that you can find MCUs that support RS-485 but lack differential transceivers? Usually this means that these MCUs facilitate communication techniques commonly used in systems with an RS-485 bus (like using nine-bit words, the USART of the AVR can do that too). The

1. In real life you may encounter several ways of labelling the two conductors. The general rule is: connect the two wires. If it does not work, swap them.

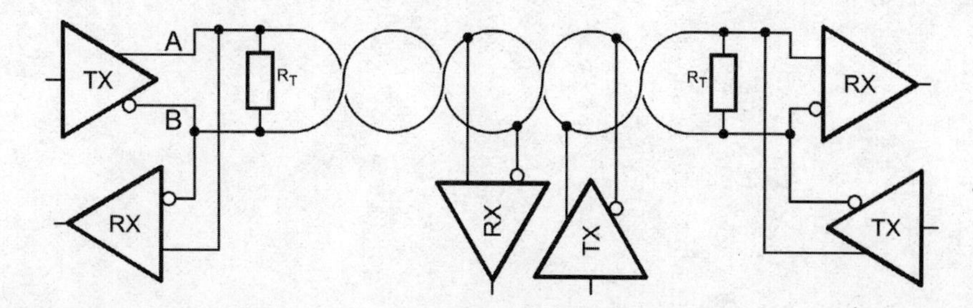

Figure 8-13 - The RS-485 bus uses differential signaling to transport data.

RS-485 standard allows networking of multiple transmitters and receivers by specifying certain driver properties. The standard does not tell you how to manage the bus or the network, or which protocol to use.

- Did you know that RS-422 is very similar to RS-485, but without the networking part?

- Did you know that the SSI (Synchronous Serial Interface) standard is to SPI what RS-422 is to RS-232, i.e. the specification of an electrical connection? It is, because SSI specifies the use of the RS-422 standard to transport the data and clock signals between a master and a slave device. However, SSI handicaps SPI because it allows data transport in only one direction, from the slave to the master.

- Did you know that even if the name I²S (Inter IC Sound) is close to I²C, the protocol is closer to that of SPI? I²S was developed for transferring audio samples between an A/D or D/A converter and a processor. It is like an SPI link between a master and a slave where the data stream never stops.

- Did you know that the MCU peripherals capable of handling SPI connections are sometimes called SSP (Synchronous Serial Port)? Such peripherals offer more flexibility than thoroughbred SPI modules, like an I²S mode or Texas Instruments' version of this type of protocol.

9. The Clock is Ticking

"*What then is time? If no one asks me, I know what it is. If I wish to explain it to him who asks, I do not know.*" [1]

Augustine of Hippo was not the only one to think about time. Man has always wondered about the nature of time and how to measure it. Historically, time was defined in terms of astronomical phenomena. The day for example, divided into hours, minutes and seconds, is defined as the time required for the earth to make one revolution on itself. The problem of measuring time was solved in 1967 when the General Conference on Weights and Measures (CGPM) redefined the second as being equal to 9,192,631,770 oscillations of a photon emitted by the transition of an electron between the two hyperfine levels of the ground state excitation of the cesium-133 atom (please excuse me if I made a mistake in the reproduction of this definition). The CGPM could just as well have chosen 9 GHz or 10 GHz or another round value as the definition of the second, but it did not. The definition of the CGPM was chosen to be as close as possible to the historic value of the second. However, the last word on the second has not yet been said. Indeed, in 2012, the American physicist David Wineland received the Nobel Prize[2] for trapping individual ions using lasers, a feat that – according to experts – is expected to change once again the definition of the second.

9.1 This is Radio Frankfurt

The existence of a definition of the second based on atoms has led to the creation of atomic clocks that provide the absolute time because they are based on the definition of the second. Owning such a clock is not really an option for the electronics enthusiast – or for any other enthusiasts for that matter – but this does not exclude their use in homemade systems. Every GPS satellite for instance is equipped with an atomic clock, and the time included in certain NMEA 0183A sentences produced by GPS receivers therefore is the absolute time[3]. It is also possible to use the time signals transmitted by radio stations such as DCF77[4] in Mainflingen, near

1. The Confessions of St. Augustine, Book 11, Chapter 14 (about 397 AD).
2. Co-winner with the French physicist Serge Haroche who has accomplished a similar feat.
3. Absolute, not exact. The distance between the transmitter and the receiver introduces a small time difference.

Frankfurt, Germany, the MSF in England or France Inter in France. In Europe the DCF77 time signal is used a lot due to the large range of the signal that covers almost all of Europe, and small and low-cost receiver modules capable of receiving this transmitter are easy to find. The way the absolute time is encoded in the DCF77 signal makes this transmitter an ideal source to begin our studies and experiments that will allow us to tame and manage time with a microcontroller.

I could now describe in detail how exactly the carrier is modulated and all that, but since DCF77 receiver modules cost only between ten and twenty dollars, I prefer to focus on the content of the data transmitted rather than the technical details of this transmission. Section 9.6 (page 236) contains some more details on the modulation of the signal.

The DCF77 time signal is pretty slow, even compared to your first internet modem with its – at the time impressive – maximum data rate of 14.4 Kbaud. Indeed, the data frames that carry the time information are sent once per minute with a bit frequency of 1 Hz (1 baud). The bits provide the seconds counter, which is not encoded in the data stream. A bit can have two durations, a logic zero lasts 100 ms, a logic one lasts 200 ms. The 59th bit (or second) is not transmitted, which allows the receiver to synchronize to the data stream. The beginning of the first bit, bit 0, is the start of a new time frame that will be valid at the beginning of the next frame.

9.1.1 DCF77

There are obviously several DCF77 receiver modules available from different manufacturers with different options, but they all offer a digital output from which we can collect the data. It is also possible to scavenge a module from a so-called "radio-controlled" clock or watch found in gadget stores and supermarkets for very little money. Some modules have an open-collector (or open-drain) output and need a pull-up resistor, others do not. The supply voltage varies from one module to another, but most operate from 5 V. Depending on the type, the receivers found in clock and watch modules probably do not support such a high voltage and you will need a level adapter to make the output signal compatible with an Arduino board. The module I use in this book is probably one of the least expensive, most common and easiest to find. This module can be powered from 5 volts and provides two outputs, one of which is inverted with respect to the other. A pull-up

4. The transmitter's acronym is not as precise as the time code it provides: D stands for Germany (but in German), C means long wave transmitter, F refers to Mainflingen (near Frankfurt) where the transmitter is located, and 77 indicates a carrier frequency of 77.5 kHz.

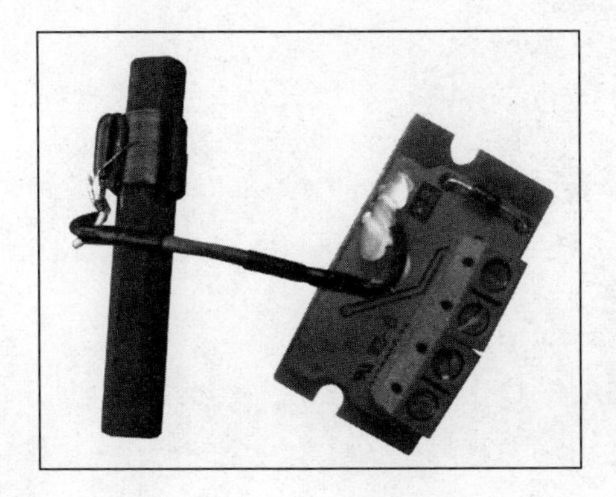

Figure 9-1 - A cheap and easy to find DCF77 receiver module. Here Frankfurt should be to the left or right of the ferrite rod.

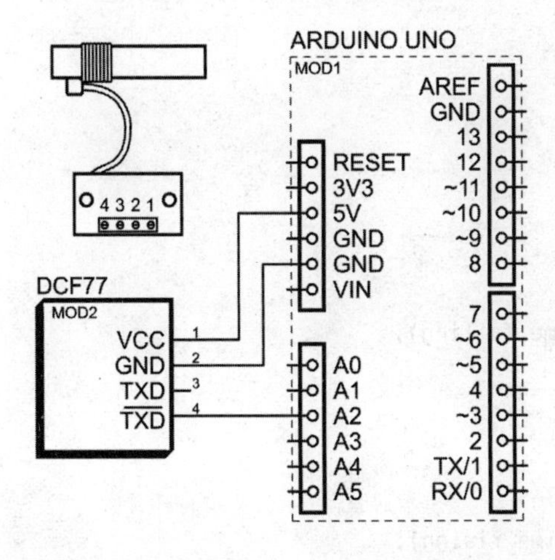

Figure 9-2 - Besides being easy to find, DCF77 receiver modules are also easy to use.

resistor is necessary; the one integrated in the MCU is perfect for the job. For the best reception orient the axis of the antenna – the ferrite rod with the coil – perpendicular to Frankfurt.

The circuit diagram shows how I connected the module to an Arduino board. I used the inverted output (pin 4), therefore my time bits are active low, but that does not change the way to process them. Since the values of the bits are encoded by time intervals, their levels do not matter. I chose analog input number two as the input for the time signal, because it was still free on my experimentation board. Here is a little sketch to verify that the module can "see" the transmitter in Mainflingen and to find out if the bits are active high or low.

```
/*
 * DCF77 polling pulse measuring
 */

int dcf77 = A2;
int pulse_prev;
unsigned long time_falling;
unsigned long time_rising;

void setup(void)
{
  Serial.begin(115200);
  pinMode(dcf77,INPUT_PULLUP);
  pulse_prev = digitalRead(dcf77);
  time_falling = millis();
  time_rising = millis();
}

void loop(void)
{
  unsigned long time = millis();
  int pulse = digitalRead(dcf77);

  if (pulse==1 && pulse_prev==0)
  {
    // rising edge
    time_rising = time;
    Serial.print("lo: ");
    Serial.println(time_rising-time_falling);
  }
  else if (pulse==0 && pulse_prev==1)
  {
    // falling edge
    time_falling = time;
    Serial.print("hi: ");
    Serial.println(time_falling-time_rising);
  }

  pulse_prev = pulse;
}
```

At each pass, the function `loop` first retrieves the number of milliseconds that have gone by since the start of the sketch. This value is provided by the function `millis`. There is a similar function called `micros`, which provides the number of microseconds since program start, but because we expect bits with a duration of at least 100 ms, a microsecond resolution offers little to no benefits and we do not need this function here.

The next step is to measure the input level and compare it to the level obtained during the previous pass. If the two levels are different, we have found an edge. If it is a rising edge, the number of milliseconds is stored in the variable `time_rising`; in case of a falling edge this number is stored in `time_falling`. The difference between the two times is calculated in accordance with the edge type and the duration obtained is sent to the serial port. Finally, the pulse level is stored for the next pass through the function `loop`.

Since the output of the receiver module needs a pull-up resistor, I configured the input as `INPUT_PULLUP` in the function `setup`.

Here is an example of the output produced by this sketch:

```
lo: 99
hi: 902
lo: 98
hi: 1900
lo: 102
hi: 898
lo: 198
hi: 806
lo: 192
hi: 809
lo: 94
hi: 903
```

This series of measurements tells us several things. The first observation to make is that the bits are active low, since only the lines that start with "lo" (short for "low") contain values that correspond to the expected bit durations. The second thing to note are the bit durations that seem not very accurate. In the list above they vary between 94 and 102 ms for a 100 ms bit and between 192 and 198 ms for a 200 ms bit. In my recording of several minutes I found values between 92 and 103 ms for a logic zero and between 190 and 198 ms for a logic one. Thirdly, we notice the line "hi: 1900" which is a frame start marker. In my multi-minute recording, I found values from 1898 to 1901 ms for this marker. It seems that the longer the interval to be measured, the more accurate the result.

9.2 Daisy-Chaining Seconds

Thanks to our test sketch, we now know that we are able to receive the DCF77 time signal. In addition, we also know the polarity of the bits and we have some proof that the function `millis` can be used to measure the bit durations with sufficient accuracy. Now it is the time to collect the bits and assemble the time frame. For this I suggest to use a `String` object. This may seem a somewhat surprising choice, but the advantage of the `String` object is that it handles the addition of new elements all by itself and that it supports the function `substring`, which will prove handy later on when we are going to decode the frame. Here is the modified function `loop` that stores the received bits in `bits`, an object of type `String`:

```
/*
 * DCF77 polling pulse measuring 2
 */

String bits;

void loop(void)
{
  unsigned long time = millis();
  unsigned long dt = 0;
  int pulse = digitalRead(dcf77);

  if (pulse==1 && pulse_prev==0)
  {
    // rising edge
    time_rising = time;
    dt = time_rising-time_falling;
  }
  else if (pulse==0 && pulse_prev==1)
  {
    // falling edge
    time_falling = time;
    dt = time_falling-time_rising;
  }
  pulse_prev = pulse;

  if (dt>50)
  {
    if (dt<150)
    {
      bits += '0';
    }
    else if (dt<250)
    {
      bits += '1';
    }
    else if (dt>1500)
    {
      Serial.println(bits);
      bits = "";
    }
  }
}
```

Instead of sending the bit durations on the serial port, they are now stored in the variable `dt`. If the value of `dt` is greater than 50 ms but less than 150 ms, we can assume that we are dealing with a logic zero and a character '0' is stored in the frame `bits`. If `dt` exceeds 150 ms, but is less than 250 ms, the value of the received bit is a logic one and the character '1' is added to `bits`. Finally, if `dt` exceeds 1500 ms, we may assume that we have detected the absence of bit number 59 that indicates the end of the frame. The frame is then sent over the serial port prior to being emptied to make room for the next frame.

With this method the polarity of the bits is irrelevant and the sketch will work with all signals, inverted or not. As you will have noticed, I chose a margin of ±50 ms for the two bit durations and even more for the end-of-frame marker. These margins may be too large and to improve your chances to detect only valid bits, you can reduce them.

The frames sent on the serial port look like this:

```
0010000
00101001001110000100100001001010010011100110000010010010000
01110000011111000100110001000010010011100110000010010010000
00001011011010100100101001000010010011100110000010010010000
```

They all contain 59 bits, except the first – which corresponds to a fragment of a frame – because the algorithm was not yet synchronized with the bit stream.

The next step is to decode the frame. The DCF77 format is well documented on the internet allowing me to fill this table with the decoded contents of the first frame of 59 bits (the second line):

Bit(s)	Description	Received	Decoded
0 to 14	Reserved	001010010011100	-
15	Transmitter is standby	0	0
16	Switch to daylight saving time at the beginning of the next hour.	0	0
17, 18	Offset from UTC.	10	2
19	Add a leap second at the end of the hour.	0	0
20	Start of time information.	1	1
21 to 27	Minutes	0000100	10
28	Parity over bits 21-27	1	1

Table 9-1 - The meaning of the bits in a DCF77 time frame (continued overleaf).

Bit(s)	Description	Received	Decoded
29 to 34	Hours	010010	12
35	Parity over bits 29-34	0	0
36 to 41	Day of the month	111001	27
42 to 44	Day of the week	100	1
45 to 49	Month	00010	8
50 to 57	Year	01001000	12
58	Parity over bits 36-57	0	0

Table 9-1 - The meaning of the bits in a DCF77 time frame (end).

From this table we learn that I received this frame on the 27th of August 2012 at 12:10 PM. The first twenty bits are not used in most applications, unless we want to know the UTC time. Bits 20, 28, 35 and 58 can be used to verify whether the data has been received correctly. Bit 20 is always one, the others are parity bits, which means that their values are zero if the number of ones in the field they monitor is even; they will be one if this number is odd. Take for example bit 35 from the table above. Because it is zero, the number of ones in the Hours field must be even, which is indeed the case.

The data of interest can be found in the fields that follow bit 20. They contain so-called Binary-Coded Decimal (BCD) values. A four-bit BCD number is similar to a hexadecimal number of four bits except that it can have only ten values, from 0 to 9. A byte can hold two four-bit BCD values. One contains the tens, the other the units. The weights of the bits are shown in the table below:

Bit No.	7	6	5	4	3	2	1	0
Weight	80	40	20	10	8	4	2	1

Table 9-2 - The weight of the bits of a binary coded decimal value.

As always, and as the table shows, bit 0 is the Least Significant Bit (LSB) and it is at the rightmost position in the table. By following this convention, the BCD byte 0100 1001 (0x49) has a decimal value of 1×40 + 1×8 + 1×1 = 49. In general, the least significant bit is the rightmost bit (and the Most Significant Bit or MSB is on the left), but in the example of the DCF77 frame, it is the leftmost bit. In the DCF77 format table above, all the fields after bit 20 have their LSB on the left[1].

1. Bits 17 and 18 contain a two-bit BCD value with its most significant bit on the left.

9.3 Decode a String of Bits

We can now write a function capable of converting a string of bits representing a BCD value into a decimal value. With such a function decoding the date and time fields of the time frame is a walk in the park.

```
/*
 * DCF77 polling pulse measuring 3
 */

int seconds;
int minutes;
int hours;
int day;
int day_of_week;
int month;
int year;

int decode_bcd(String str)
{
  unsigned int i;
  int power_of_two;
  int result = 0;
  int power_of_ten = 1;
  for (i=0; i<str.length(); i++)
  {
    power_of_two = 1<<(i&0x3);
    result += (str[i]-'0') * power_of_two * power_of_ten;
    if (power_of_two==8) power_of_ten *= 10;
  }
  return result;
}

boolean decode_time(String &str)
{
  if (verify(str)==false) return false;
  minutes = decode_bcd(bits.substring(21,28));
  hours = decode_bcd(bits.substring(29,35));
  day = decode_bcd(bits.substring(36,42));
  day_of_week = decode_bcd(bits.substring(42,45));
  month = decode_bcd(bits.substring(45,50));
  year = decode_bcd(bits.substring(50,58));
  return true;
}
```

The `decode_time` function simply calls the `decode_bcd` function for each field following bit 20 of the time frame. Note how we make use of the function `substring` of the `bits` object – which is a `String` object – to extract effortlessly the fields to decode. The function `verify` that is called at the beginning of `decode_time` will be explained below. The function `decode_bcd` is somewhat more complex. In the loop of this function, all the bits in the subframe `str` are multiplied by their respective weights and the results are added. The weight of the bits can be decomposed into a power of ten and a power of two (for example, $40 = 2^2 \times 10^1$, see also the table above). The power of two that we need at each iteration can be derived from the two least significant bits of the loop counter `i` – extracted with the bitwise AND operator '&' – that take on the periodic binary sequence 00, 01, 10, 11, 00, 01, etc. or, in decimal: 0, 1, 2, 3, 0, 1, etc. The power of two is obtained by using this two-bit value to shift the number one up to three places to the left, which creates the powers of two 1, 2, 4 and 8 (you may want to write them in binary notation to better understand what is happening). Because the LSB is the first of the bits in the string, the power of two starts at 1, as does the power of ten. Whenever the power of two reaches the value of 8, the power of ten is updated by multiplying it by ten. This way we get the sequence 1, 2, 4, 8, 10, 20, 40, 80, 100, 200, etc. The bits are actually characters in `str` ('1' and '0'), not numbers, so the character '0' is subtracted from each bit. This trick takes advantage from the fact that in ASCII encoding the character '0' has a value of 0x30 and the character '1' equals 0x31: '0' – '0' = 0x30 – 0x30 = 0 and '1' – '0' = 0x31 – 0x30 = 1 (see also Section 7.3.6, page 138).

Phew, that was a lot of words to explain three lines of code. Let's conclude with a remark. The loop counter `i` is an unsigned integer (**`unsigned int`**). The reason for this is that the value returned by the function `String.length` is also unsigned and if `i` had been a signed type, the compiler would have issued a warning informing you that you were trying to assign an unsigned value to a signed variable. Remember, we strive to always eliminate all compiler warnings, which is why I made `i` an unsigned integer.

9.3.1 DCF77 Decoder

Now let's check the time frame using its parity bits. Here are two functions that together can fulfil this important task:

```
boolean parity(String str, char parity_bit)
{
  unsigned int i;
  int sum = 0;
  for (i=0; i<str.length(); i++)
  {
    if (str[i]=='1') sum += 1;
  }
  return ((sum&1)+0x30 == parity_bit);
```

```
}

boolean verify(String str)
{
  if (str.length()!=59) return false;
  if (str[20]!='1') return false;
  if (parity(str.substring(21,28),str[28])==false) return false;
  if (parity(str.substring(29,35),str[35])==false) return false;
  if (parity(str.substring(36,58),str[58])==false) return false;
  return true;
}
```

The function `verify` checks everything it can without understanding the contents of the frame. The first check is to see if the length of the frame is equal to 59, not more or less. A second test concerns bit 20 that must always have the value of one. The function then continuous by verifying the parity bits of the fields Minutes and Hours and of the compound date field. The first test that fails immediately ends the verification procedure and a negative result will be returned.

The parity bits are tested by the function `parity` with as arguments the field to be checked and the value of the parity bit as it was found in the frame. This function simply counts all the ones present in the field (in Section 9.6.1, page 239, you will see another method to calculate the parity). The final test is the most complex line. In binary representation bit 0 of an even number is always equal to zero, for an odd number bit 0 is always one. Bit 0 is extracted with a bitwise AND before being transformed into an ASCII character by adding the value 0x30 (or the character '0', see also above to perform the inverse operation). The character obtained must be equal to the parity bit (character), if it isn't there is a parity error. Remember, if the number of ones is odd, the parity bit must be equal to one to create an even number. If the number of ones is even, the parity bit must be zero to keep the number of ones even.

We can now have a look at the function `loop` that allows us to send the decoded time to the computer. The only thing that changes over the previous version is the call of the function `decode_time` and of course the transmission of the decoded values to the computer. I also added a seconds counter which unfortunately does not work well in the Arduino serial monitor. The idea is that the seconds are always displayed at the beginning of the line, but the monitor does not properly handle the carriage return (`Serial.write(0xd)`). The method works perfectly fine in a more serious serial port terminal program (such as the free but a bit complicated TeraTerm).

```
void loop(void)
{
  unsigned long time = millis();
  unsigned long dt = 0;
  int pulse = digitalRead(dcf77);

  if (pulse==1 && pulse_prev==0)
  {
    // rising edge
    time_rising = time;
    dt = time_rising-time_falling;
    seconds += 1;
  }
  else if (pulse==0 && pulse_prev==1)
  {
    // falling edge
    time_falling = time;
    dt = time_falling-time_rising;
  }
  pulse_prev = pulse;

  if (dt>80)
  {
    Serial.write(0x0d);
    Serial.print(seconds);
    if (seconds<10) Serial.print(' ');

    if (dt<120)
    {
      bits += '0';
    }
    else if (dt>180 && dt<220)
    {
      bits += '1';
    }
    else if (dt>1800)
    {
      Serial.println();
      if (decode_time(bits)==true)
      {
        Serial.print(hours);
        Serial.print(':');
        Serial.println(minutes);
        Serial.print(day);
        Serial.print('/');
        Serial.print(month);
        Serial.print("/20");
        Serial.println(year);
      }
      else Serial.println("error");
      bits = "";
```

```
      seconds = 0;
    }
  }
}
```

I also tightened the detection margins for the duration of the bits a little. Together with parity testing, decoding of the DCF77 time frame has become quite robust. To further enhance the robustness, you can add checks to the decoded values. The hours may not exceed 23; the minutes should not exceed 59, etc. I leave this exercise to you.

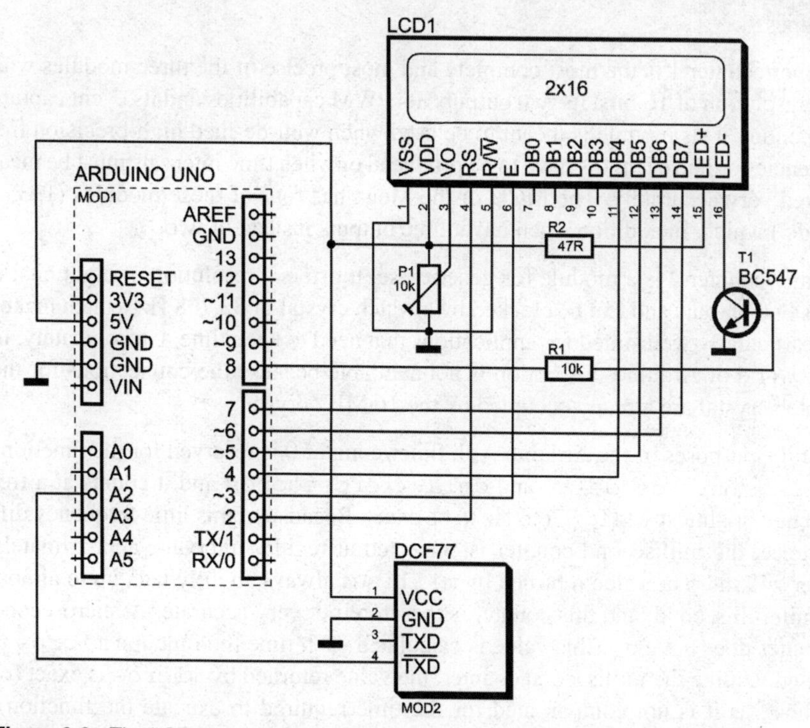

Figure 9-3 - The LCD shows the time obtained from the data provided by the DCF77 receiver module.

9.4 Millis and Micros, Two Little Functions

Writing the sketch to decode the DCF77 time frame was relatively easy thanks to the function `millis` that allowed us to measure the duration of the bits without much effort. This function uses one of the timer/counter modules of the microcontroller to count the number of milliseconds since the start of the sketch. The MCU on the Arduino Uno board has three of these modules, the one on the Mega board has six. The modules are not all identical. Timer/counter 0 is an 8-bit module, sporting two independent outputs and offering PWM functionality. It was intended by its designer to be used in synchronization tasks and to produce rectangular wave shapes.

Timer/counter 1 is the most complete and most precise of the three modules with its resolution of 16 bits, its two outputs, its PWM capabilities and its event capture functions. This module is meant to be used when well-defined high-precision frequencies or pulse lengths have to be produced or when time intervals must be measured very accurately. The MCU of the Mega has four of these modules (1, 3, 4 and 5) which, in addition, each have three outputs instead of two.

Timer/counter 2 is a module for generic use. It offers a resolution of eight bits; it has two outputs and can be clocked by a watch crystal of 32,768 Hz, which makes it particularly well suited for applications that need to keep time. Unfortunately, in the world of Arduino this option is not available because the connections for the watch crystal are already occupied by the 16 MHz crystal.

For the purposes of the Arduino API, timer/counter 0 is reserved for the functions `millis`, `micros`, `delay` and `delayMicroseconds` and it counts at a frequency of almost 1 kHz, 976.6 Hz to be exact. Because of this little frequency difference, the millisecond counter is corrected at regular intervals, approximately every 42 ms. The value returned by `millis` is always slightly late, up to almost a full millisecond, and this counter is therefore not very accurate. A microsecond counter does not exist, this value is calculated each time the function `micros` is called. Unlike the millisecond counter, the value returned by `micros` is exact (or almost, as it is not compensated for the time required to execute the function). Given the inaccuracy of `millis`, the function `delay` uses `micros`. The function `delayMicroseconds` performs its own calculations and depends on no other functions. It is fairly accurate, except for very small delays of only one or two microseconds.

9.5 PWM

All the counters/timers of the MCU on the Uno have two outputs and they all know how to do PWM, which explains the six PWM outputs that are available for `analogWrite`. The MCU on the Mega does not have sixteen of these outputs as you might think, but fifteen, as the supplementary – third – output of timer/counter 1 (compared to the Uno) shares a pin with the second output of timer/counter 0. Note that the Arduino documentation is lagging a bit behind the ever evolving Arduino hardware and API, and an update would be much appreciated. At the time of writing, according to the Arduino reference, the Mega board still has only twelve PWM outputs (which was the case for the previous version of the Mega).

9.5.1 Two Types of PWM

The quality of the PWM signals produced by `analogWrite` is supposed to be independent of the output used; the signal should be the same no matter which PWM output you activate. However, in reality this is not the case. Although the counter/timer modules all work in 8-bit mode with a clock of 16 MHz / 64 = 250 kHz (a prescaler value of 64), they do not generate the same frequency. Pins 5 and 6 that are connected to timer/counter 0 run at almost 977 Hz (16 MHz / 64 / 256), the other outputs deliver a frequency of just over 490 Hz (16 MHz / 64 / 255 / 2). The PWM signals on pins 5 and 6 (pins 4 and 13 on the Mega) are faster but less symmetrical than the signals on the other outputs that make use of an option of the MCU called phase correct PWM. Timer/counter 0 on the other hand operates in fast PWM mode. This difference is due to the fact that timer/counter 0 is also used as milliseconds counter. If you use timer/counter 0 in phase correct PWM mode, which is quite possible, the milliseconds counter will stop working correctly, the values provided by the functions `millis` and `micros` will be wrong and the function `delay` will be too slow.

To produce a PWM signal, the counter is compared to a threshold value. If the counter is below this threshold, the PWM signal is low; if the counter exceeds the threshold, the PWM signal is high[1]. In fast PWM mode the counter counts from 0 to 255 and then starts over, because 255 + 1 = 0 in 8-bit-register arithmetic. The counter clock is therefore divided by 256. The waveform generated by the counter is a sawtooth and the pulses of the PWM signal are always to the left of the maximum counter value. In contrast, in phase correct PWM mode, the counter will not continue at 0 after reaching the maximum value, but it will instead change direction. The value following 255 is 254, not 0. Whenever the counter reaches the

1. The levels may be inverted.

maximum or minimum value, it changes direction. The waveform generated by the counter is triangular in this case, with two ramps instead of one. Consequently, the counter crosses the threshold value twice, once going up and once counting down, with the result that the pulses of the PWM signal are now centred on the maximum counter value. The frequency of this signal is slightly higher than half of the fast PWM signal's frequency, because the clock counter is in this case divided by 255 × 2. To visualize the difference between the two PWM methods, compare with an oscilloscope the two outputs of the same timer/counter module while they produce two different duty cycles.

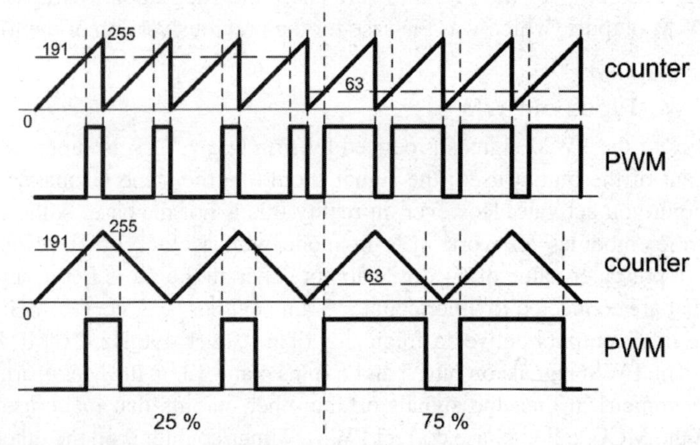

Figure 9-4 - The difference between fast PWM (top) and phase correct PWM (bottom).

9.6 The Master of Time

Apart from `analogWrite`, the Arduino API does not offer other possibilities to play with PWM signals. This is unfortunate because the manufacturer of the MCU is quite proud of the PWM capacities of the AVR. So, to highlight the efforts of the manufacturer, I will present here a spectacular and unique application (let's remain modest): a PWM-based DCF77 time signal transmitter.

Radio-controlled clocks synchronize themselves to the DCF77 time signal (or to the signal of another time signal station) received over air. The time signals from the various transmitters, not only the DCF77 time code, are so poorly protected and their formats are so well documented that they are easy targets for malevolent people. As a matter of fact, it is not too difficult to build your own compatible time signal transmitter and disturb radio-controlled clocks nearby. Doing so they will show the time you want, instead of the real time. The use of GPS or mobile phone jammers is legal under certain conditions, so why wouldn't a time signal jammer be legal? And besides that, it's simply fun. So, how can we accomplish this?

The signal transmitted by the DCF77 transmitter is quite simple. It is composed of a carrier, a 77.5 kHz sinusoidal signal, modulated in amplitude by the time information[1]. To send a bit the amplitude of the carrier is reduced to 20% of the non-modulated value. The duration of the modulation determines the bit value. A logic zero has a duration of 100 ms, a logic one lasts twice as long, 200 ms. Several methods can be employed to produce such a signal, for instance with an accurate sine oscillator followed by an attenuator with two levels (20% and 100%), controlled by a microcontroller which provides the time code bits. Producing a frequency of 77.5 kHz is quite within the reach of an MCU and it would be a shame not to take advantage of this capability. The MCU of the Arduino board is clocked at 16 MHz. If we divide this frequency by 206 we obtain a frequency of 77,670 Hz, an error of only 0.2% with respect to 77.5 kHz. Cheap commercial DCF77 receivers are probably not picky enough to make a fuss about such a small difference.

The next step is to apply the amplitude modulation (AM) to the carrier. The solution that comes to mind is a voltage divider with two resistors, but that requires an additional MCU pin to control the divider. However, it is possible to keep things even simpler by using pulse width modulation. By doing a "little" mathematics, Fourier analysis in this case, it is possible to demonstrate that each periodic signal can be constructed from a series of sinusoidal signals. The first sinusoidal signal in the series, the one with the lowest frequency, is called the fundamental, and its frequency is the same as that of the periodic signal. The other sinusoids are the harmonics. Their frequencies are integer multiples of the fundamental frequency, and their amplitudes and phases depend on the waveform of the periodic signal. The series of sine waves form what is called the spectrum of the periodic signal and, like the fingerprint of humans, it is unique for each waveform[2]. As an example, a square wave contains only harmonics with frequencies that are odd multiples of the fundamental frequency. We can for instance construct a square wave from sine waves with frequencies of 77.5 kHz, 232.5 kHz (x 3), 387.5 kHz (x 5) and so on.

---

1. In fact, the format is not that simple, because the signal is partly phase modulated by a pseudo-random sequence to enable the correction of the received time for the distance between the transmitter and the receiver. Fortunately for us, the vast majority of commercial receivers does not make use of this information.
2. If the average value of the periodic signal is not zero, the spectrum will also contain a DC component (0 Hz).

When the duty cycle of the square wave deviates from 50%, even harmonics also appear such as 155 kHz (× 2) and 310 kHz (× 4). The amplitude distribution also changes. Not only do the amplitudes of the harmonics depend on the duty cycle, the amplitude of the fundamental also is a function of this parameter. It is at its maximum for a duty cycle of 50% and diminishes when the duty cycle decreases or increases. We can use this property to modulate the amplitude of the fundamental. For our DCF77 jammer all we have to do is choose a rectangular signal with a duty cycle different from 50% allowing the amplitude of its fundamental to be equal one-fifth of the amplitude of the fundamental of a square wave.

What duty cycle should you choose? If you are good at math, you can calculate it; if you are not, like me, you can use your multimeter or oscilloscope. There is another solution: get yourself a Fourier analysis tool (there are many on the internet) and try several signals to obtain the desired result. I used the oscilloscope method.

A filter tuned to the fundamental frequency is used to extract a lovely 77.5 kHz amplitude-modulated sine wave from a PWM signal. This filter should be selective without being world-class (although that doesn't hurt). It should filter out all the harmonics starting from the first at 155 kHz. Amplitude modulation also changes the average signal level, the DC level, and a band-pass filter which by its nature eliminates the average level, is the best choice. I got good results with an active filter of the fourth order. With a multimeter or oscilloscope connected to the output of the filter you can measure the amplitude of the fundamental for different duty cycles and thus find the correct value.

The last hardware step is to equip the output stage with an amplifier and an antenna. The frequency on which most atomic clocks transmit is so low that it takes a powerful amplifier and a huge antenna to obtain a usable range. The wavelength λ at 77.5 kHz is 3871 m and the optimum $\lambda/4$ antenna is therefore almost 968 meters (3200 ft.) long. Even a less efficient $\lambda/20$ antenna still requires a cable, wire or conductor of a length of almost 200 m (600 ft.). Such antennas can be constructed by winding the conductor around a frame to create a frame (or rhombic; window) antenna. I chose a different approach, much less fatiguing, because I simply cannibalized the antenna of a DCF77 receiver module. It is a ferrite rod antenna that does not support high power because the ferrite saturates quickly, but that has the considerable advantage of being already tuned to the right frequency with a capacitor soldered on the coil, and by the positioning of the coil on the ferrite rod. OK, the matching of this antenna with the MCU output will be far from optimal, but it is good enough for these experiments. With this antenna, I managed to get a range of about two meters (7 ft.) with an identical DCF77 receiver module. For a

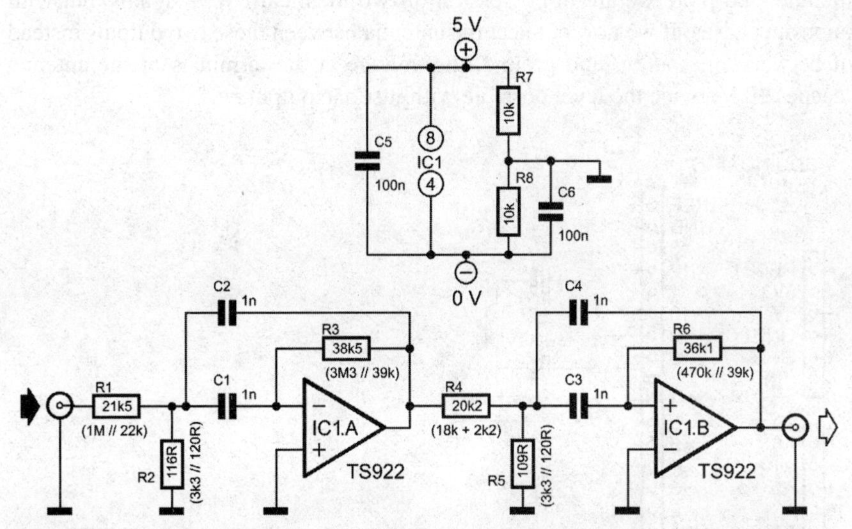

Figure 9-5 - A fourth order Bessel band-pass filter allows you to view the fundamental of a PWM signal on an oscilloscope. This filter has been calculated for a center frequency of 77.5 kHz. The resistor values in parentheses are used to make the non-standard values from standard "E12" values in series or in parallel. Beware of the asymmetrical power supply.

cheap commercial clock the range was about 20 cm (almost 8 inches). To improve the range an amplifier capable of producing tens of watts and a good transmitting antenna are needed, but their use is probably illegal.

9.6.1 DCF77 Transmitter

For my final design I did not use a filter or an amplifier. Since the transmitting antenna is an LC filter, and the receiver has a similar antenna, and since Mother Nature tends to attenuate the high frequencies of a signal, the harmonics will not have that much influence on the signal. As a result, the role of the amplifier could be endorsed by the MCU and I could connect the antenna in series with a capacitor – to block the DC voltage, which in this setup has no use whatsoever – to pin 9 of the Arduino board.

A way to improve the range of a transmitter is to increase its output voltage. This seems impossible without an external amplifier, but by involving the second channel of timer/counter 1 it can be done. Channel B is identical to channel A, so it is possible to produce exactly the same signal on output B. To invert the signal on this output all we have to do is set some bits in the configuration register of the module, and there we are in possession of two identical PWM signals, but with opposite phases. If we now connect the antenna between these two outputs instead of between one output and ground, the voltage at the terminals of the antenna reaches 10 V, twice the level possible with just one output.

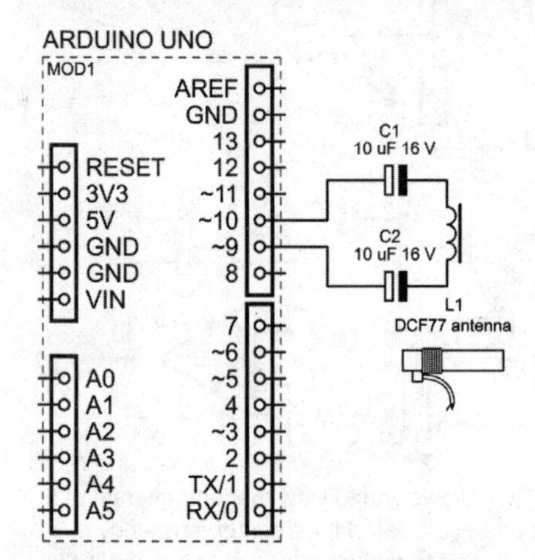

Figure 9-6 - World's simplest DCF77 transmitter.

Now let's move on to the sketch that will produce the PWM-based DCF77 time signal.

As explained above, we cannot use the function `analogWrite` to generate a 77.5 kHz PWM signal. There is also the function `tone` but its frequency does not exceed 65,535 Hz (the argument to pass the frequency is limited to sixteen bits) and furthermore it cannot do PWM. Yet the AVR is capable of generating frequencies up to half the MCU's clock frequency, which corresponds to 8 MHz in the case of Arduino. Luckily the API is only an API and we do not have to limit ourselves to the functions it offers. However, bypassing it requires further study of the technical specifications of the MCU.

The timer/counter modules have seven registers in common. Here we are only interested in three of these registers: TCCRxA, TCCRxB and OCRxA where you should replace the 'x' with the module's number. The TCCR (Timer/Counter Control Register) registers determine the way the module operates. The TCCRxA register connects the output pins and defines how to use them. TCCRxB controls the clock for the module, its source and the value of the prescaler. These two registers share a field named Waveform Generation Mode (WGM) which allows you to choose what the output will look like and what its maximum frequency will be. This field has four bits for 16-bit modules and three bits for 8-bit modules. The table in the datasheet that explains this field is as clear as mud. It is so incomprehensible that I'm not even going to try to clarify it here. Instead, I'll explain my reasoning to figure out the correct mode.

A frequency of 77.5 kHz for a PWM signal is quite fast. If we allow 256 duty cycle values, the maximum rate will be 16 MHz / 256 = 62.5 kHz, too low for our application. We need to divide by 16 MHz / 77.5 kHz = 206, and we need a fairly good PWM resolution to allow 80% amplitude modulation of the fundamental. Phase correct PWM divides the maximum frequency by two, so we have to choose 8-bit fast PWM. The counter may not exceed 206, our maximum or TOP value as the datasheet calls it, which means that we cannot use the built-in TOP constant of 255 (0x00ff in the datasheet); we need to use a register to set the maximum value (which also determines the signal's frequency). Now we are left with the choice between using the OCRxA register to set the maximum value – an option offered by all modules – or using the ICR register, an option available only on 16-bit modules. But beware, there is a snag…

The OCR (Output Compare Register) registers are used to specify the counter values (thresholds) above which the outputs change level (they control the duty cycle). When the counter's maximum – or TOP – value is the built-in constant 255, that's fine, but when you have to use the same register to specify the maximum value **as well as** the duty cycle, there is something wrong. Indeed, this is not possible. The datasheet talks about PWM, but when we choose OCRxA to set the maximum value, the duty cycle is either 0%, 50% or 100%, which is not very flexible. In short, this option is a non-option (but can be used to generate an 8 MHz square wave).

By elimination we now arrive at the conclusion that we will have to use the ICR (Input Capture Register) register to specify the maximum counter value. Then the OCRxA register will be available to specify the duty cycle. Only 16-bit timer/counter modules have this register, and the MCU of the Uno has only one of them: timer/counter 1. So now we know which timer/counter to use (no. 1) and which PWM mode (no. 14). This also fixes the pins to use: 9 and 10. Pin 10 (channel B) will be used to produce the same signal as that supplied by pin 9, but with opposite phase.

Here is the sketch that puts into practice what you just have read:

```
/*
 * DCF77 transmitter
 */

int output = 9;
```

```
int output_inverted = 10;
int led = 13;
int frequency = 206;
int am = 10;
String bits;
uint8_t utc_offset = 2;
uint8_t seconds = 0;
uint8_t minutes = 40;
uint8_t hours = 17;
uint8_t day_of_week = 4;
uint8_t day_of_month = 6;
uint8_t month = 9;
uint8_t year = 12;

uint8_t days_in_month[12] =
{
  31, 28, 31, 30, 31, 30, 31, 31, 30, 31, 30, 31
};

uint8_t to_bcd(uint8_t value)
{
  // the maximum value we handle is 99.
  if (value>99) value = 99;
  uint8_t tens = value/10;
  uint8_t units = value - tens*10;
  return (tens<<4) | units;
}

uint8_t sentence_add_value(uint8_t value, uint8_t nr_of_bits,
                                                        String &str)
{
  uint8_t i;
  uint8_t parity = 0;
  uint8_t bcd_value = to_bcd(value);
  for (i=0; i<nr_of_bits; i++)
  {
    if ((bcd_value&1)!=0)
    {
      str += '1';
      parity ^= 1;
    }
    else str += '0';
    bcd_value >>= 1;
  }
  return parity;
}

void sentence_build(String &str)
{
  char parity;
  str = "00000000000000000";                         // clear bits 0 to 16
```

```
  if (utc_offset==2) str += "10";
  else str += "01";
  str += "01"; // add fixed bits 19 & 20
  parity = sentence_add_value(minutes,7,str);
  str += char('0'+parity);
  parity = sentence_add_value(hours,6,str);
  str += char('0'+parity);
  parity = sentence_add_value(day_of_month,6,str);
  parity ^= sentence_add_value(day_of_week,3,str);
  parity ^= sentence_add_value(month,5,str);
  parity ^= sentence_add_value(year,8,str);
  str += char('0'+parity);
}

void time_tick(void)
{
  seconds += 1;
  if (seconds>59)
  {
    seconds = 0;
    minutes += 1;
    if (minutes>59)
    {
      minutes = 0;
      hours += 1;
      if (hours>23)
      {
        hours = 0;
        day_of_week += 1;
        if (day_of_week>7) day_of_week = 1;
        day_of_month += 1;
        if (day_of_month>days_in_month[month])
        {
          // Don't handle leap years.
          day_of_month = 1;
          month += 1;
          if (month>12)
          {
            month = 1;
            year += 1;
          }
        }
      }
    }
  }
}

void setup(void)
{
  pinMode(output,OUTPUT);
  pinMode(output_inverted,OUTPUT);
```

```
  // Counter1, PWM mode 14, prescale of 1, OC1A normal, OC1B inverted
  TCCR1A = _BV(COM1A1) | _BV(COM1B1) | _BV(COM1B0) | _BV(WGM11);
  TCCR1B = _BV(WGM13) | _BV(WGM12) | _BV(CS10);
  ICR1 = frequency;
  OCR1A = ICR1 >> 1;              // duty cycle 50%
  OCR1B = OCR1A;

  seconds = 0;
}

void loop(void)
{
  sentence_build(bits);

  uint8_t bit_length = 0;
  if (seconds<59)
  {
    bit_length = 100 << (bits[seconds]-'0');
    OCR1A = am;                   // 20% of maximum amplitude
    OCR1B = OCR1A;
    delay(bit_length);
    OCR1A = ICR1 >> 1;            // amplitude 100% (duty cycle 50%)
    OCR1B = OCR1A;
  }
  delay(1000-bit_length);

  time_tick();
}
```

As before, we start reading at the highest level in the function `loop`. Here we begin by constructing the string of time bits to transmit. It is assembled from the global time and date variables `year`, `month`, `day_of_month`, `day_of_week`, `hours`, `minutes`, `second` and `utc_offset`. As in the DCF77 receiver sketch, the frame `bits` is an object of type String.

Remember, the end of the time frame is indicated by the absence of a pulse, which is why we check if the seconds counter is less than 59 and not equal (the counter starts at 0). If a bit is to be sent, we need to determine its duration. To show that I am good at programming, I did this in an obfuscated way. It would have been much clearer to write

```
if (bits[seconds]=='1') bit_length = 200;
else bit_length = 100;
```

but that occupies two lines (or any other inexcusable reason). Measure the time you need to understand my single line of code, and then never ever do as I did. A real good programmer will always try to write comprehensible code.

To send a bit we should reduce the amplitude of the fundamental to about 20% of its maximum value. To do this, the duty cycle of the PWM signal is modified using the constant `am` (short for amplitude modulation) whose value was obtained during the experiences above, and the sketch waits `bit_length` milliseconds before the duty cycle is restored to 50% (maximum amplitude). Note how channel B always follows channel A.

The function `loop` now waits the remaining interval of 800 or 900 ms to complete the second and then ends with a call to `time_tick` to advance the sketch's time and date one second.

In the function `setup` you can see how to initialize the timer/counter 1 in mode 14 with a clock prescaler value of 1. In this mode the ICR1 register controls the frequency of the PWM signal while OCR1A (and OCR1B) controls its duty cycle. It is also important to note that we need to configure pin 9 and 10 as outputs with a call to `pinMode`. If you forget to do this, which is easy, there will be no output signal.

Remains to explain the construction of the time frame and how to advance the sketches time and date. Let's start at the end, with the function `time_tick`. This function is so simple that you will be able to understand it without my help. That said, I did leave you the exercise to add leap year correction.

Building the time frame begins in the function `sentence_build` that concatenates the different fields using the function `sentence_add_value`. The latter – supported by the function `to_bcd` – not only converts a value to a string of zeroes and ones, but it also calculates the parity of the string. This time it is done with the exclusive OR (EXOR) '^' operator of C. The variable `parity` is an integer that is converted to an ASCII character by adding the character '0'. The typecast with the **`char()`** function here does not convert the variables into **`char`** types – they are already of this type – but circumvents a problem due to the interpretation of the type of the result of the addition of '0' and `parity`. Here this result is interpreted as an integer, not a character, and without a typecast it will be converted into a string of two characters ('4' and '8' or '9', depending on the value of `parity`) instead of a character '0' or '1' (remember, the ASCII character '0' has a value of 48 decimal, '1' has a value of 49). The frame would then contain four illegal characters, and, worse, it would have had two characters too many (61 instead of 59). By putting the addition in a typecast, all goes well.

9.7 Could do Better

The limited range of the DCF77 transmitter described above is a bit disappointing. You can make a window antenna, but that is a bit tedious. And what if we injected the time signal into the home powerlines? In a building, powerline wires go everywhere, they are long, they are invisible, but, thanks to wall outlets and lamp sockets, they are accessible and it is easy to create a loop wherever you want. Such a loop around a receiver is very effective indeed. So, how can we achieve this?

In DIY shops and supermarkets you can find cheap AC powerline intercoms that use a square wave of about 100 kHz as a signal carrier, not that far at all from the DCF77's 77.5 kHz carrier frequency. The powerline interface in my intercom is controlled by a simple NPN transistor. Therefore I disconnected the powerline interface from the rest of the intercom, and I connected pin 9 of the Arduino board to the base of the transistor. Then I connected the assembly to the powerline in the garage, near the electric switchboard. The result was impressive. My receiver placed in another room close to a power outlet or a lamp received the signal without any problems. A power cord looped around a commercial battery powered radio-controlled clock allowed me to tamper with it. This technique should work even better with an AC-powered clock, but I have not tried this because I did not have one. Imagine that you tamper with your spouse's radio-controlled alarm clock this way. After forcing him or her to oversleep several times – nothing beats a repeated prank – you risk finding yourself alone again, allowing you to spend all your free time on programing microcontrollers.

One final remark on this subject before we continue with measuring time with a microcontroller. For the sake of saving energy, commercial radio-controlled clocks and watches do not listen continuously to the time code transmitter. Since the signal is weak, it has to be amplified to exploit it and that takes energy. In addition, the signal is slow, and you have to listen a long time to capture a complete time frame. Therefore, in order to make the battery last as long as possible, these clocks only check the time occasionally, once every hour or even once a day; the rest of the time they rely on their own crystal. So if you want to know if a radio-controlled clock receives your homemade time code signal well, you must either be very patient or force the clock to resynchronize. They usually have a small button which allows you to do this. After pressing this button the clock will try for a few minutes – again to save energy – to receive the time signal. After this period, if the clock managed to decipher the time code it will set itself to the received time. If it did not receive anything usable, it will simply assume that it is on time. In both cases it will now cut the power to its receiver. This tells us two things. First of all,

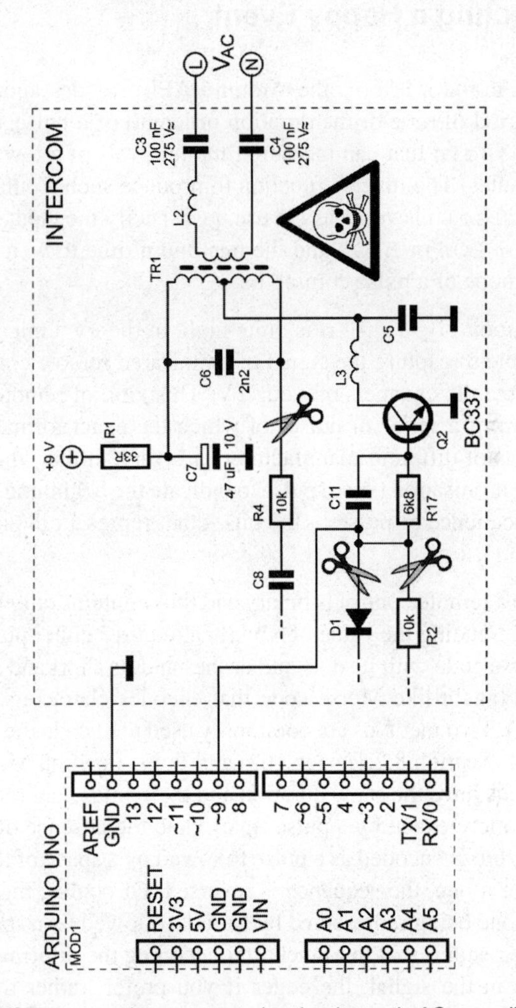

Figure 9-7 - The DCF77 transmitter connected to the domestic AC powerlines. Always be very careful when you connect something to the AC grid.

a little patience is needed to determine if your setup is working and what its range is. Secondly, running a clock in reverse or freezing its hands is not possible with these types of clocks; all you can do is change the time they show.

9.8 Expecting a Happy Event

Besides `millis` and `micros` the Arduino API provides another function for measuring a period of time or the duration or length of a pulse. I am referring to the function `pulseIn` that can measure time intervals of a few microseconds to about three minutes. The inverse function to produce such a pulse does not exist. The function `pulseIn` is very easy to use, just specify the input pin, the expected level of the pulse `LOW` or `HIGH` and the maximum time to wait (timeout) before abandoning all hope of a pulse coming in.

Since the resolution of `pulseIn` is quite high, in theory a microsecond, we can use it for example to capture the signal of an infrared remote control like the one you use to change the channels on your TV. This kind of remote control sends a code in the form of a series of pulses of which the exact format depends on the brand. The codes of different manufacturers are very similar; they all begin with at least one synchronisation (sync) pulse to indicate the beginning of the bit stream, followed by a sequence of pulses – the bits – that represent the information to be transmitted.

The signal from a remote control is binary and thus contains only two states: signal on or signal off. Sending the value of a bit then requires either pulse length modulation (like Morse code with its dots and dashes and its short and long breaks) or a pulse sequence (again like Morse code that encodes characters as sequences of dots and dashes). Two methods are commonly used to encode the bits: Manchester coding[1] (see also Section 8.2.1, page 162) and Pulse-Position Modulation (PPM). In the first, all bits have the same duration and their values are represented by two pulses, or to be more exact, by a pulse (mark) and the absence of a pulse (space). For example, a zero is encoded as a pulse followed by a space of the same duration as the pulse. For a one, the sequence is reversed. Of course, the opposite is also possible with a one being represented by a mark followed by a space and a zero as the inverse. The result is that in Manchester encoding the information is conveyed by the changes in the signal, the edges if you prefer, rather than by the pulse lengths. If a rising edge indicates a one, a falling edge indicates a zero and vice versa. The average value of a Manchester-encoded signal is constant (50%).

1. *Differential* Manchester encoding (note the word differential) is the same as bi-phase encoding. In bi-phase encoding the bits all have the same duration. The end of each bit is indicated by a level change. One of the bit values, generally the '1', contains an additional level change at half the pulse length. A sequence with only zeroes produces a square wave with a frequency of f Hz, a sequence that contains only ones will produce a square wave with a frequency of $2 \times f$ Hz.

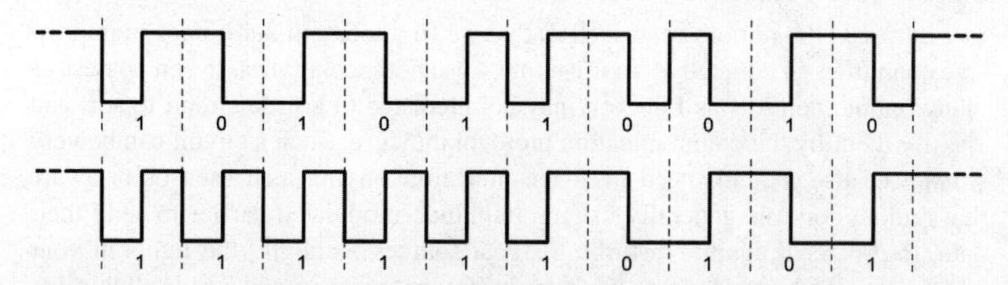

Figure 9-8 - Two examples of Manchester encoding (here a space is shown as a high level). Since the information is stored in the level changes, the number of pulses needed to transmit a byte depends on the value contained in the byte. The length of a byte is always the same.

When the duration of a bit depends on its value we talk about pulse-position modulation or PPM. There are several ways to do this:

1. The mark length is constant; the space length is variable. A zero is shorter than a one.

2. The mark length is constant; the space length is variable. A one is shorter than a zero.

3. The mark length is variable; the space length is constant. A zero is shorter than a one.

4. The mark length is variable; the space length is constant. A one is shorter than a zero.

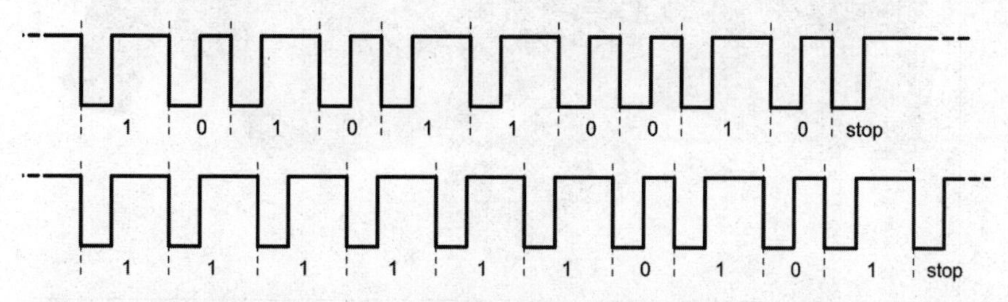

Figure 9-9 - Two examples of PPM encoding (here a space is shown as a high level). Since the information is stored in the bit lengths, the transmission of a byte always contains the same number of pulses. The length of the frame depends on the value contained in the byte. A stop pulse is necessary to signal the end of the frame.

9.8.1 Sort Your Infrared Remote Controls

The infrared (IR) remote control (RC) that we find today in our homes in impressive quantities – I counted 18 in my home – use both techniques. If you possess as many remote controls as I do you may be interested in a little circuit to sort and classify them by the communication protocol they use. Such a circuit can be very simple because you only need an IR sensor. Before buying such a sensor, be aware that remote controls generally use an amplitude-modulated carrier to send their data, to protect it against disturbing signal sources (sunlight, the lights in your home, etc.). This carrier often has a frequency between 36 and 38 kHz, but other frequencies are also possible.

Demodulating this signal is not too complicated, but it is even easier to use a sensor with integrated demodulator. One manufacturer who specializes in this area is Vishay, and they offer a wide range of infrared sensors.

Figure 9-10 - Can you find the intruder hiding among these remotes?

For my experiments I first used a once very popular TSOP1736 found in my junk box, but it is now obsolete. I then redid my experiences with a TSOP34836. The number 36 in its reference means that the sensor is optimized for a 36 kHz carrier, but it also captures remotes that transmit on a frequency nearby. This part worked perfectly fine with all my remotes.

The sensor has three connections: power, common and signal. In my case, the signal is Active Low and provides pulse frames as in the illustrations above, that is to say that in the case of a space (no signal) the output will be at 5 V. The component consumes only a tiny amount of current and can be powered by an output of the MCU. This makes it suitable for connecting directly to an extension connector of the Arduino board without anything else to wire. Real Plug'n'Play.

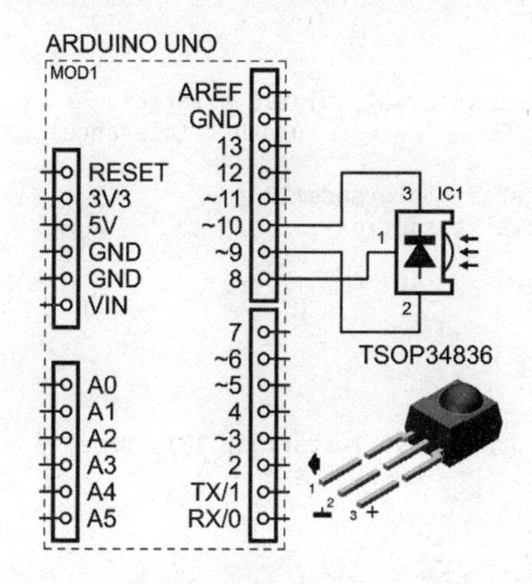

Figure 9-11 - Only few things are easier than connecting an infrared remote control receiver chip to an Arduino board.

The sketch to sort your remote controls is slightly more complicated than the circuit. Here it is:

```
/*
 * guess IR format
 */

int ir_signal = 8;
int ir_vcc = 10;
int ir_gnd = 9;

typedef struct
{
  uint16_t startbit;            // in microseconds
  char name[10];
```

```
}
protocol_t;

#define PROTOCOLS_MAX  10
const protocol_t protocols[PROTOCOLS_MAX] =
{
  {  140, "RECS80" },
  {  275, "DENON" },
  {  512, "MOTOROLA" },
  {  890, "RC5" },
  { 2400, "SIRCS" },
  { 2685, "RC6" },
  { 3380, "JAPANESE" },
  { 4500, "SAMSUNG" },
  { 8000, "DAEWOO" },
  { 9000, "NEC" },
};

boolean interval_within_bounds(uint16_t interval, uint16_t target,
                                                    uint8_t tolerance)
{
  unsigned long dx = (unsigned long)target*tolerance/100;
  return (interval>target-dx && interval<target+dx);
}

void decode(unsigned long pulse)
{
  for (int i=0; i<PROTOCOLS_MAX; i++)
  {
    if (interval_within_bounds(pulse,protocols[i].startbit,10)==true)
    {
      Serial.print(protocols[i].name);
      break;
    }
  }
  Serial.print(" - ");
  Serial.println(pulse);
}

void setup(void)
{
  Serial.begin(115200);
  pinMode(ir_signal,INPUT);
  pinMode(ir_gnd,OUTPUT);
  digitalWrite(ir_gnd,LOW);
  pinMode(ir_vcc,OUTPUT);
  digitalWrite(ir_vcc,HIGH);
  // Suppress spurious pulses on power up.
  delay(500);
  while (pulseIn(ir_signal,LOW,500000)!=0);
}
```

```
void loop(void)
{
  unsigned long pulse = pulseIn(ir_signal,LOW,10000000);
  if (pulse!=0)
  {
    decode(pulse);
    while (1);
  }
}
```

The function `loop` begins with a call to the function `pulseIn`. As arguments we pass the number of the pin that is connected to the sensor's output, the expected level of the pulse (active low in my case) and the time window within which the pulse must present itself to be taken into account. I set it to ten seconds, leaving you plenty of time to launch the sketch, grab the remote, aim it at the sensor and press one of the RC's buttons.

If no pulse is detected within ten seconds after the call to `pulseIn`, this function will return zero and a new detection window is opened. When a pulse is detected `pulseIn` measures its duration and returns the value found in microseconds. This value is almost exact, and it is passed without corrections or modifications to the function `decode` before entering an infinite loop. To restart the sketch you should therefore press the reset button.

As usual, the function `setup` contains nothing very exciting. The sensor is powered by the pins 9 and 10. I sequenced the sensor's power supply by applying the 0 V before the 5 V. A delay of 500 ms allows the sensor to get used to its supply. Despite this precaution, I still observed spurious pulses during a few milliseconds on the signal output of the sensor. Capturing them with `pulseIn` in a loop allowed me to get rid of them. This loop terminates when no pulse is detected for a period of 500 ms. Therefore you need to wait at least one second after launching the sketch before pressing a button on the remote.

The function `decode` compares the measured pulse length to the theoretical durations stored in a table. Since remote controls are not very accurate, the measured duration rarely corresponds perfectly to a theoretical duration and we therefore apply a tolerance of 10% using the function `interval_within_bounds`. If a match is found in the table, the name of the protocol is sent to the serial port, followed by the measured pulse length itself. If the function `decode` cannot identify the pulse, only the pulse length is sent on the serial port.

9.9 Break or Continue

You probably have noticed the C keyword **break** in the function decode. This instruction allows a program to (prematurely) exit a loop. The **break** statement is the complement of the **continue** statement that skips the rest of the loop to continue with the next iteration. In other words, the **break** statement causes the MCU to continue the program just after the loop's closing curly brace whereas **continue** makes the MCU jump back to just before the **for**, **while** or **do** of the loop.

```
while (1)
{
  if (the_burner_is_off==true)
  {
    break;
  }
  else if (the_milk_does_not_boil_yet==true)
  {
    continue;
  }
  it_is_boiling_over();
  do_something();
}
```

This example shows an infinite loop and the only way out of it is to turn off the gas (if possible when the milk begins to boil or before). The **break** statement will then make the MCU jump out of the loop. If the milk does not simmer yet, everything is fine and the last instructions are skipped thanks to the **continue** statement. If however you were distracted and the milk moved beyond the simmer stage while you replied to a text message, it will boil over and you will have some cleaning up to do. Turning off the gas always gets you out of this loop.

9.10 Divide and Conquer

Did you notice the typecast of the variable target in function interval_within_bounds? It is necessary to avoid a possible overflow in the calculation that without this typecast will be executed in 16-bit arithmetic. The order of operations is important in this calculation, because the variables are all integers and cover a fairly wide range. If we divide before multiplying, we risk losing too many decimal places. For example, the smallest value in the table is 140, which, divided by 100, gives 1. Thus, the tolerance that will be applied will be

$1 \times 10 / 140 = 7\%$ instead of 10%. The largest value in the table is 9,000 and dividing it by 100 is not a problem.
On the other hand, $9{,}000 \times 10 = 90{,}000$, which is too large to be stored in a 16-bit variable. The variable will overflow and all that remains is 24,464[1]. The tolerance that will be applied is now $24{,}464 / 100 / 9{,}000 = 27\%$. The cast of the calculation to **long** (32 bits) avoids this problem. In calculations it is important to keep the intermediate results as large as possible. Also try to keep the divisions for the end. It's like in electronics, where we always try to keep the Signal to Noise Ratio (SNR) as high as possible.

9.11 The Structured Union of Types

The previous sketch introduced two more new elements of the C language: the definition of a data type and the structure to group several variables. Let's start with the structure.

9.11.1 struct

The last sketch contained three pin declarations for the infrared sensor. These three parameters are part of the same object, the sensor. If there had been two or more identical sensors in this circuit, the list of pin declarations would have been longer, making it easy to get lost or to make mistakes elsewhere in the program. A method for grouping variables that belong together like the sensor pins will be welcome. The C keyword **struct** provides exactly that. It allows you to create a structure of several variables of any type. Here is how you declare a structure for three variables (of the same type, but that is accidental):

```
struct
{
  int gnd;
  int vcc;
  int signal;
}
infrared_receiver;
```

1. To understand this, perform the calculation in hexadecimal notation:
0x2328 × 0xa = 0x15f90. Only the sixteen least significant bits fit into a 16-bit word. The result is therefore truncated and becomes 0x5f90, or 24,464 in decimal.

In the remote control data format detector sketch above, I have defined a structure to easily create a table of correspondence between start pulse durations and the names of the protocols. To access the elements of a structure, you have to write the name of the structure followed by the name of the element, the two separated by a dot, exactly like when you use the serial port or an object of type `String`:

```
infrared_receiver.gnd = 9;
infrared_receiver.vcc = 10;
infrared_receiver.signal = 8;
```

9.11.2 union

There is a second type of structure in C, the **union**. The big difference between the **struct** and the **union** is the memory footprint. All the elements of a **struct** have their own memory location, and the example above occupies six bytes: two per integer. If you replace in this example the keyword **struct** by the keyword **union** – which is totally legal – then the three integers will occupy only two bytes. Strange, isn't it? Not really, once you know that in this case you can only store a single integer in the three parameters. So the three variables offer three ways to access the same data. A **union** is used to group data that can represent several things at once. For example:

```
union
{
  int arms;
  int legs;
  int ears;
  int lungs;
}
human;
```

A normal person has two of each.

9.11.3 typedef

The C keyword **typedef** allows you to define a data type. Just write **typedef** followed by the type definition. The Arduino types like `boolean`, `byte` and `word` are defined this way, but they merely redefine another type. For example, `boolean` is a redefinition of `uint8_t`, which is a redefinition of **unsigned char**. This is more data type renaming than defining a new one. What is more interesting is to define compound types of several variables:

```
typedef struct
{
  char last_name[10];
  char first_name[10];
```

```
  unsigned char age;
  boolean runny_nose;
}
child_t;
```

To avoid confusion with a variable, I always use the suffix "_t" to indicate that this is a definition of a type.

A data type defined with **typedef** is used like any other type:

```
void wipe_your_nose(child_t kid)
{
  kid.runny_nose = false;
}

child_t snotty = { "Valens", "Thibaut", 10, true };
wipe_your_nose(snotty);
```

Note how the fields of the type can be initialized in one fell swoop, but without forgetting the curly braces.

Defining your own data types can improve greatly the readability of a program; it allows you to pass several variables in a single function argument or to easily adapt a data type for another platform. If, for example, you use in your program variables of type boolean that occupy one byte on platform A and two bytes on platform B, just change the definition of boolean to adapt your program to the platform used.

9.12 Is It an Image? Is It data? It's Superfile!

Detecting the start bit of a frame transmitted by a remote control is one thing, decoding all the bits of the frame is something else. It is often helpful to visualize the received data prior to thinking up an algorithm for decoding it. This is of course possible with a (digital) storage oscilloscope but there are other methods for those who do not have such a tool. In Section 7.3.2 (page 129) I showed you how to use the MCU to produce data that can be loaded into a spreadsheet or be written to a so-called Comma Separated Values (CSV) file. This is a technique often used by data loggers.

9.12.1 The SVG File Format

To view remote control data frames I propose another technique, based on the Scalable Vector Graphics (SVG) file format. Such a file can be viewed in a web browser or with a modern image processing program (like the free tool Inkscape, www.inkscape.org). The SVG format is interesting for us, because it consists only of eXtensible Mark-up Language (XML) instructions, which are human readable and easy to teach to an MCU. "Drawing" in such a file is equivalent to writing a list of (x,y) coordinates and specifying a pencil. No need to calculate pixels or colors or to fill out complicated headers. Here is a simple little SVG program that draws the letter 'A' in a browser (put it in a file that you save as a `.svg`, then open the file in the brower) :

```
<svg xmlns="http://www.w3.org/2000/svg" version="1.1">
<polyline points="0,100 25,0 42,70 8,70 42,70 50,100"
style="fill:white;stroke:black;stroke-width:5"/>
</svg>
```

The first line is the header that prepares the browser for the data that follows. The next two lines contain the drawing and the last line ends the program. The drawing consists of a broken line – indicated by the instruction `polyline` – that passes through the *x,y* coordinates following the keyword `point`. The origin (0,0) is at the top left of the screen, thus the drawing is somewhat upside down. How the broken line must be drawn is specified by the `style` statement: `fill` specifies the background color, `stroke` specifies the line color and `stroke-width` the width of the line. Whitespace and line breaks do not affect the interpretation of the drawing.

How can we use this type of file to visualize the data sent by a remote control? We know that our data consists of variable length pulses that only take on two levels. With the function `pulseIn` we can measure the pulse lengths, so we can draw a broken line for each pulse: a vertical line segment downwards of free length, a horizontal segment to the right of which the length is a function of the pulse duration, then a vertical line upwards with the same length as the first vertical line.

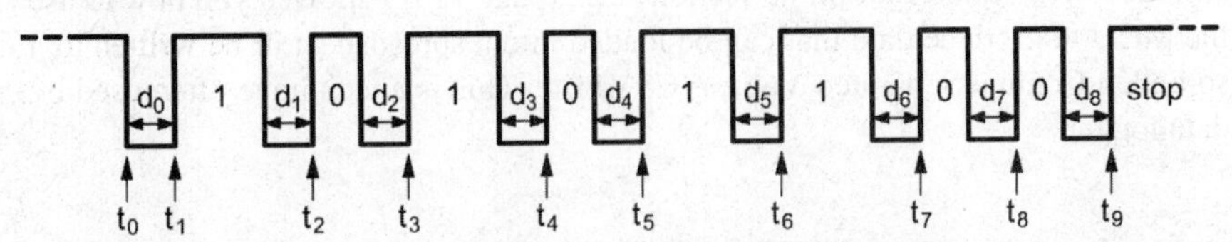

Figure 9-12 - Reconstructing an RC data frame from a set of duration measurements (d_n) and timestamps (t_n).

The information that is missing to complete the drawing are the lengths of the spaces that separate the pulses. Fortunately it is easy to get these by using the API function `micros` that counts the time elapsed since the start of the sketch in microseconds. If we call `micros` right after the return of `pulseIn`, we know the time when the pulse ended. By subtracting the pulse length from it we obtain the start time of the pulse. In short, by recording for each pulse the time when the function `pulseIn` returns together with the duration of the pulse, we have enough information to reconstruct the frame and we can draw the missing horizontal line segments between the pulses. Here is the result:

```
/*
 * export IR pulses to SVG file
 */

#define X_SCALE  25

int ir_signal = 8;
int ir_vcc = 10;
int ir_gnd = 9;
unsigned long time_start;
int x_offset = 10 * X_SCALE;

typedef struct
{
  unsigned long timestamp;
  unsigned long pulse;
}
point_t;

#define POINTS_MAX  50
point_t points[POINTS_MAX];
int index;

void svg_write_file(int x_offset, int x_scale, int y_min, int y_max)
{
  Serial.print("<svg xmlns=\"http://www.w3.org/2000/svg\"
                                                  version=\"1.1\">");
  Serial.print("<polyline points=\"0,");
  Serial.print(y_max);
  Serial.println(' ');
  for (int i=0; i<index; i++)
  {
    Serial.print((points[i].timestamp-points[i].pulse)/x_scale);
    Serial.print(',');
    Serial.print(y_max);
    Serial.print(' ');
    Serial.print((points[i].timestamp-points[i].pulse)/x_scale);
    Serial.print(',');
```

```
    Serial.print(y_min);
    Serial.print(' ');
    Serial.print(points[i].timestamp/x_scale);
    Serial.print(',');
    Serial.print(y_min);
    Serial.print(' ');
    Serial.print(points[i].timestamp/x_scale);
    Serial.print(',');
    Serial.println(y_max);
  }
  Serial.println("\" style=\"fill:white;stroke:black;stroke-width:1\"
                                                            /></svg>");
}

void setup(void)
{
  Serial.begin(115200);
  pinMode(ir_signal,INPUT);
  pinMode(ir_gnd,OUTPUT);
  digitalWrite(ir_gnd,LOW);
  pinMode(ir_vcc,OUTPUT);
  digitalWrite(ir_vcc,HIGH);
  // Suppress spurious pulses on power up.
  delay(500);
  while (pulseIn(ir_signal,LOW,500000)!=0);
  time_start = 0;
  index = 0;
}

void loop(void)
{
  unsigned long pulse = pulseIn(ir_signal,LOW,2000000);
  if (pulse!=0 && index<POINTS_MAX)
  {
    unsigned long time_now = micros();
    if (time_start==0)
    {
      time_start = time_now - pulse - x_offset;
    }
    points[index].timestamp = time_now - time_start;
    points[index].pulse = pulse;
    index += 1;
  }
  else
  {
    svg_write_file(x_offset,X_SCALE,150,50);
    while (1);
  }
}
```

As in the previous sketch, the function `loop` begins with a call to the function `pulseIn`. If it returns a value greater than zero, a pulse was detected and we store the timestamp (in microseconds) in the variable `time_now`. The pair (`time_now`, `pulse`) is stored in an array of points that we will process once we have received the entire frame. The end of the frame is reached either when `pulseIn` did not detect a pulse within a period of two seconds, or when the table is full. When the sampling period is over, the points are drawn in an SVG file and the sketch ends in an endless loop. Press the board's reset button to restart the sketch.

Of course, the sketch does not really write a file, it only sends the contents of a file. It is your job to copy the data from the serial monitor into a file and save it with the extension SVG.

You have probably noticed the variable `time_start` used to correct `time_now`. Remember, the function `micros` returns the number of microseconds since the launch of the sketch, a value of more than one million since the sketch already waited for one second in the function `setup` (as it did also in the previous example). To avoid that the first pulse is drawn at position 1,000,000 or more to the right, we correct the values provided by the variable `micros` for this initial period. This correction is stored in the variable `time_start`. I added a variable `x_offset` so that it easy to move the beginning of the broken line.

The table where the coordinates are stored is an array with elements of type `point_t`, a data type that I defined myself at the top of the sketch using a **`typedef`**. It is a **`struct`** with two elements of type **`unsigned long`**. This structure occupies eight bytes, therefore I limited the size of the array to 50 elements. You can save a lot of memory here if you use elements of type **`unsigned int`**, because the pulse duration is probably always less than 65,535 μs (the maximum value for this variable).

The function `setup` is identical to the one in the previous sketch, except for the initialization of two additional variables.

The third function `svg_write_file` creates the SVG file that contains the drawing of the frame. Since whitespace and line breaks do not matter, I cut the instruction `polyline` into several pieces so that it is easier to write the point coordinates.

Note how the horizontal axis is scaled by the variable `x_scale`. This is necessary to view the entire frame in the browser. The points that fall off the screen (which at the time of writing generally have a width of 1,280 pixels) are not displayed. For a reason unknown to me current internet browsers cannot scroll SVG drawings. An image processing program should do this a little better.

Surely you have noticed the forward slash '\' characters that appear neither in the SVG file example above nor in the data displayed on the serial monitor. They are due to quotation marks that are found all over the contents of an SVG file. In C the quote is reserved for delimiting strings, so you have to use a trick to store a quote in a string. This trick is the so-called escape sequence that consists of adding a reserved character ('\') in front of another reserved character to transform the latter into a normal character. Thus you have to write "\"" to print the quote. This technique also allows you to write non-printable characters like the value zero ("\0", not to be confused with the character '0') or the tabulation character ("\t")[1]. Printing the forward slash itself is done by repeating it once (i.e. "\\").

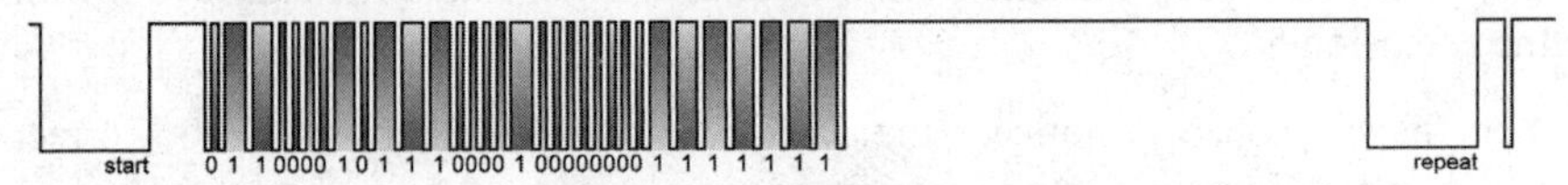

Figure 9-13 - An IR RC frame in NEC-1 format captured in an SVG file. It was me who colored the bits so nicely to improve their visibility.

The SVG file format not only allows us to easily create charts and curves in for instance a web browser, but it can also store the raw data. The SVG file can thus be both a graph and a data recording at the same time. I think we will hear much more about this format in the future.

9.13 What They Really Say

Since there are many manufacturers of remote controls, there are also many remote control communication protocols. According to the literature (the internet and electronics magazines) a common format in Europe would be the Philips protocol RC5 – that, by the way, uses Manchester encoding – and many circuits using it can be found. However, among all my eighteen remotes I have found only one that that actually used the RC5 format. Publications employing the SIRCS format from Sony can be found too, but I do not have a compatible remote control. I do have lots of remote controls that use the NEC-1 format from NEC (now Renesas) or something similar to it. This is hardly surprising when we take into account that NEC is a Japanese company and most Asian gadget manufacturers shop in Asia.

1. Try also "\7", that should make some noise.

Since NEC protocols seem to be so popular, I decided to show you how to decode it with an Arduino board. Once you master this, you can start concocting home-made codes and disrupt TV sets, Hi-Fi systems and other remote controlled devices around you.

9.13.1 The NEC-1 Protocol

The NEC-1 protocol uses the first variant of PPM encoding where all pulses have the same duration, but the length of the spaces is variable, and a zero is shorter than a one. To decode such a frame, you need to lock onto the pulse stream, and measure the time between pulses. The pulse length itself is not interesting. Officially, a pulse indicates the beginning of a bit, but we can also consider a pulse as the end of the previous bit. Since the NEC-1 frame size ends with a Stop pulse, we can use this second interpretation here.

```
/*
 * decode NEC IR RC protocol
 */

#define TIMEOUT  150000
#define PULSES_MAX  34
#define TOLERANCE  15  /* % */
```

```
int ir_signal = 8;
int ir_vcc = 10;
int ir_gnd = 9;

typedef struct
{
  unsigned long timestamp;
  unsigned long on_time;
}
point_t;

point_t pulses[PULSES_MAX];
int index;

// En microsecondes
uint16_t startbit = 9000;
uint16_t on_time_0 = 560;
uint16_t off_time_0 = 565;
uint16_t on_time_1 = 560;
uint16_t off_time_1 = 1690;
unsigned long time_start;
unsigned long rc_code = 0;

boolean interval_within_bounds(uint16_t interval, uint16_t target)
{
  unsigned long dx = (unsigned long)target*TOLERANCE/100;
  return (interval>target-dx && interval<target+dx);
}

boolean bit_add(uint8_t value, uint16_t interval, uint16_t target)
{
  if (interval_within_bounds(interval,target)==true)
  {
    rc_code <<= 1;
    if (value!=0) rc_code |= 1;
    return true;
  }
  return false;
}

boolean decode_nec(void)
{
  int i;
  uint16_t interval;
  unsigned long t_prev = pulses[0].timestamp;
  uint8_t pulsecount = PULSES_MAX;

  rc_code = 0;
  for (i=1; i<pulsecount; i++)
  {
    // 4 intervals possible: normal bits, bit 1 & repeat code.
```

```
    interval = pulses[i].timestamp - t_prev;
    if (interval_within_bounds(interval,on_time_0+startbit/2)==true)
    {
      // first data bit.
      if (interval_within_bounds(pulses[i].on_time,on_time_0)==false)
      {
        goto decode_nec_error;
      }
    }
    else if (interval_within_bounds(interval,on_time_0
                                                +startbit/4)==true)
    {
      // repeat code.
      if (interval_within_bounds(pulses[i].on_time,on_time_0)==false)
      {
        goto decode_nec_error;
      }
      pulsecount = 2;
    }
    else if
           (interval_within_bounds(interval,on_time_0+off_time_0)==true)
    {
      // previous bit was a zero.
      if (bit_add(0,pulses[i].on_time,on_time_0)==false)
      {
        goto decode_nec_error;
      }
    }
    else if
           (interval_within_bounds(interval,on_time_1+off_time_1)==true)
    {
      // previous bit was a one.
      if (bit_add(1,pulses[i].on_time,on_time_1)==false)
      {
        goto decode_nec_error;
      }
    }
    else
    {
        goto decode_nec_error;
    }
    t_prev = pulses[i].timestamp;
  }
  return true;

decode_nec_error:
  Serial.print("bit ");
  Serial.print(i);
  Serial.print(", interval=");
  Serial.print(interval);
  Serial.print(", pulse=");
```

```
  Serial.print(pulses[i].on_time);
  Serial.print(", rc_code=");
  Serial.println(rc_code,HEX);
  return false;
}

boolean decode(boolean startbit_only)
{
  if (interval_within_bounds(pulses[0].on_time,startbit)==true)
  {
    if (startbit_only==true)
    {
      return true;
    }
    else
    {
      return decode_nec();
    }
  }
  return false;
}

void setup(void)
{
  Serial.begin(115200);
  pinMode(ir_signal,INPUT);
  pinMode(ir_gnd,OUTPUT);
  digitalWrite(ir_gnd,LOW);
  pinMode(ir_vcc,OUTPUT);
  digitalWrite(ir_vcc,HIGH);
  // Suppress spurious pulses on power up.
  delay(500);
  while (pulseIn(ir_signal,LOW,500000)!=0);
}

void loop(void)
{
  unsigned long pulse = pulseIn(ir_signal,LOW,TIMEOUT);
  if (pulse!=0)
  {
    unsigned long time_now = micros();
    if (time_start==0)
    {
      time_start = time_now;
    }
    if (index<=PULSES_MAX)
    {
      pulses[index].timestamp = time_now - time_start;
      pulses[index].on_time = pulse;
      if (index==0 && decode(true)==false)
      {
```

```
        // resynchronise
        time_start = 0;
      }
      else index += 1;
    }
  }
  else
  {
    if (index>0)
    {
      if (decode(false)==true)
      {
        Serial.println(rc_code,HEX);
      }
      index = 0;
    }
    time_start = 0;
  }
}
```

This sketch largely resembles the previous two. The function `setup` has not changed, and the table where the sample data is stored is the same as is the function `interval_within_bounds`. The function `loop` has been modified slightly with respect to the previous one. A simple mechanism for synchronizing to the start bit has been added to ensure that the recording begins only after receiving a start bit. Furthermore, the work that has to be done after the reception of a complete frame – in the body of the **`else`** statement at the end of `loop` – is different this time, since we now decode a frame instead of drawing it. The function `decode` plays the main role in both modifications. Its argument changes depending on the task at hand: `true` if it just has to decode a start bit or `false` when it must decode an entire frame. In this case it subcontracts the job to the function `decode_nec`. This is where it gets interesting.

When you look closely at the timing diagram of a NEC-1 frame, no doubt you can distinguish three space lengths that the decoder should recognize: those of the data bits '0' and '1', and the space between the start bit (the first pulse) and the beginning of the first bit (the second pulse). However, there is a fourth space that occurs when a key is pressed, the so-called repeat code. In this case, probably to save some energy, instead of periodically sending the same frame, the remote sends a special frame consisting of only two pulses and that does not contain data. The time between these two pulses is different from the other spaces.

As explained above, the algorithm considers that a pulse signals the end of a bit, so it calculates the interval between the pulse and the previous one to determine the value of the bit. Since the first two pulses concern the Start bit, they are only used to check the validity of the frame. The Stop pulse allows decoding of the last bit. Here's what it I got for the keys 0 to 9 of an LG television's AKB72915207 remote control:

```
20DF08F7
20DF8877
20DF48B7
20DFC837
20DF28D7
20DFA857
20DF6897
20DFE817
20DF18E7
20DF9867
```

The first four characters are the manufacturer's code (LG? I have not been able to verify this with a similar remote from LG), the next two characters contain a 4-bit address or function code and four bits of data. The last two characters contain a kind of checksum. If you add the fifth and the seventh character, the result always have to be 0xf (15), *ditto* for the sixth and the eighth character. In other words, the last byte (remember, a byte contains two hexadecimal characters) is the inverse of the second-to-last. This method not only allows you to verify the correct reception of the data, but it also ensures a constant frame duration (the manufacturer code is constant).

In this list, if we read the address character in binary from left to right, we obtain the values of 0 to 9, which corresponds perfectly to the numbers printed on the keys. The data character always contains 8. If we read it the same way as the address, we get 1, which probably means something like "key pressed".

The function `decode_nec` looks more complicated than it really is. This is due to the error detection that I included. The pulses stored in the table and the calculated intervals are checked to see if they respect the allowed tolerances. If an error is detected, the decoding stops and an error message is printed. I could have implemented this test using a special error function, but I took the opportunity to introduce yet another keyword of the C language: **`goto`**.

9.14 To goto Or Not to goto

The **`goto`** statement is one of the reasons why computer scientists have created high-level languages like C and C++. With a **`goto`** you can jump in a program from A to B without ever returning to A. You will easily understand that such a statement in the hands of a novice (like you) can be very dangerous, so I kept it for the end of the chapter (and almost the end of the book). The **goto** is a statement not to use.

The **goto** has its roots in assembly languages where jumping around is the only way to create conditional program execution.[1] A program is a list of instructions that gets executed from top to bottom (generally from bottom to top) and if a certain part of this list should not be executed because of a condition, you cannot do much more than jump over the obstacle. The first versions of the BASIC language featured the **goto** and made it very popular. Although convenient, the **goto** statement is also at the origin of the term spaghetti code which refers to a program that has become unreadable because of branches and jumps in every direction. The more structured programming languages try to force the programmer to use functions instead of **goto** statements.[2]

Having said that, sometimes the use of **goto** can be justified, as in this sketch where it allows you to easily report errors and process them all in the same way, a handy option during the development of a program. In the function `decode_nec` the **goto** statement terminates the decoding prematurely because an error was encountered. It would of course have been possible to make the function return false instead of using **goto**, but I wanted to know why the decoding failed without putting `Serial.print` instructions everywhere.

The **goto** statement is always followed by a destination label that must exist elsewhere in the program, before or after the **goto** statement. The label at the destination is in turn followed by a colon ':'. If you really need to use a **goto** statement somewhere in your program, make sure that the destination label is not too far away, otherwise you risk finding yourself in an Italian kitchen.

```
void park_the_car(void)
{
  if (no_parking_space_here==true) goto further_away;
  park_here();
  return;

further_away:
  take_the_bus();
}
```

1. In assembler language, jumps are called jumps and branches, not **goto**. Assembly languages also provide the ability to call functions so you do not have to do everything with branches.
2. Modern BASIC dialects allow programs to be written without using the **goto** statement.

9.15 Frame It Yourself

Now that we know how to capture remote control commands, we can turn our attention to transmitting our own commands. This really is not complicated, especially if you are not too afraid to delve into the internals of the Arduino API. But first things first, let's start with the hardware.

By basing the design on a microcontroller an infrared remote control can be constructed with only a few additional parts. The MCU can accomplish a large part of the work by producing and modulating the infrared carrier signal. It's a bit like our DCF77 transmitter, we just need to add an "antenna", which in this case boils down to an infrared LED connected to an output of the MCU. To avoid damaging the output we limit the current through the LED to 20 mA with the help of a resistor. I scavenged the infrared LED from an old remote of which the receiving end, a television set, was no longer in my possession. With a multimeter I measured a forward voltage of about 1 V, the current limiting resistor then should be equal to 5 – 1 / 0.02 = 200 Ω. I used 220 Ω which is a standard value.

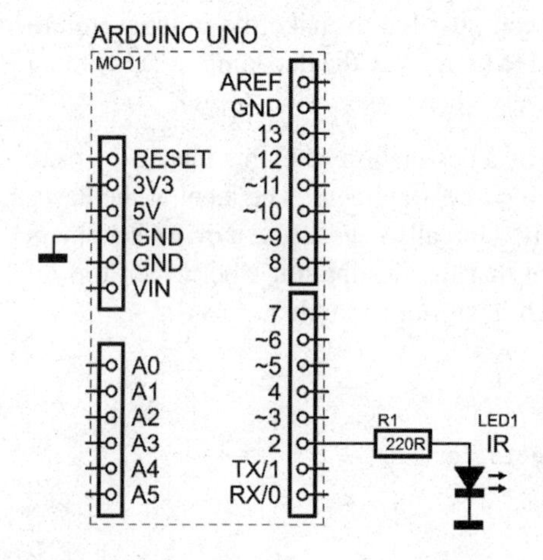

Figure 9-14 - The infrared transmitter is as simple as the receiver. Remote controlled devices are surprisingly sensitive, but if you need a longer range just add a transistor to increase the current in the LED.

9.15.1 Composition

The bits of the code to be transmitted must modulate a carrier. Each bit consists of a mark and a space. A mark corresponds to a carrier modulation of 0%, while a space is obtained with a modulation of 100%. Thus, a mark produces a pulse train with a frequency equal to that of the carrier, whereas a space only generates a time delay. Thus, to send a bit we need a function to produce a pulse train and another one capable of producing a delay. Delays are in the millisecond range, and the API's function `delayMicroseconds` is perfect for this job. What remains is the creation of the pulse trains.

The carrier frequency is usually somewhere in the range of 36 to 38 kHz, which is within the capabilities of the API function `tone` that we already used in Chapter 7 to build the Misophone. This function also allows you to specify a tone duration in milliseconds, which is unfortunate because our pulse trains will have durations of hundreds of microseconds. Therefore a exploratory tour of the function `tone` could be interesting to see if there happens to be a trick to get around the duration problem, or just to get some ideas that would allow us to write our own function `tone`.

The function of interest is located almost halfway down the file `Tone.cpp` in the subdirectory hardware\arduino\cores\arduino\ of the Arduino installation. The length of the function is mainly due to the calculation that converts the specified frequency into values for the prescaler and the OCR register (see also Section 9.6.1, page 239). Another reason for this length is that the function has been designed to work with any of the MCU's timer/counter modules, but in reality it is restricted to timer/counter 2. Therefore the variable `_timer` always contains the value of 2, a crucial detail that allows us to skip half of the code.

What interests us first is to know how exactly the function uses its duration parameter. We find it about two-thirds down the function where it is used to calculate the number of level changes needed (`toggle_count`) to generate a pulse train at the right frequency with the specified length. We could change this calculation to make it use microseconds instead of milliseconds, but that is not a good idea because it will affect all the sketches that use the function `tone` with a duration. Never change an API without ensuring that it will not have a negative impact on other programs that rely on it.[1] Let's first look closely at how the `toggle_count` variable is used, perhaps another solution exists?

A little further down, the function the variable `toggle_count` is copied into `timer2_toggle_count`, a variable managed by the function `ISR`. The file contains six copies of this function, but only the one with the argument `TIMER2_COMPA_vect` is enabled, the others are ignored because of the fact that `tone` always uses timer/counter 2. The function `ISR` is a special function, called each time counter 2 reaches the value specified by the value in the OCR register. Functions of this type will be treated in the next chapter, so I will not go into details here. All that matters now is to understand that the function `ISR` flips the level of

1. Too bad for the sketch that relies on a bug or a "feature" of the API to work.

the pin specified for the function tone and that it decrements the variable timer2_toggle_count. When the latter reaches zero, the function ISR terminates the function tone.

Now we're getting somewhere. If we modify the value of the variable timer2_toggle_count immediately after calling the function tone we can influence the length of the pulse train generated by this function. We must do this as quickly as possible after the call of tone because we need fairly short durations.

Now that we have found a way to use tone with durations of less than one millisecond, we have to put it into practice. By default we do not have access to the timer2_toggle_count variable, because it is private to the file Tone.cpp. We can make it public by changing the API, a solution that we don't like, but luckily in C there is another method of explaining to the compiler that a variable or a function exists somewhere outside our sketch. This method is the keyword **extern**. Putting this keyword in front of a definition or declaration tells the compiler that the body, the implementation, of this definition is located in another file of the project. The compiler does not need to know where the body is, its photograph (its definition) is enough. The linker collects the compiled bodies when it builds the executable file. For each missing body it will throw an error of the type undefined reference (see also Section 5.2, page 60).

Now on to the sketch which sends a code in NEC-1 format that corresponds to the '1' key on the remote control of my Orange internet TV decoder:

```
#define CARRIER  38000
#define RC_CODE  0x6170807F

int ir_led = 2;
uint16_t startbit = 9000;
uint16_t on_time_0 = 560;
uint16_t off_time_0 = 565;
uint16_t on_time_1 = 560;
uint16_t off_time_1 = 1690;
extern volatile long timer2_toggle_count;

void ir_send_pulse(uint8_t pin, uint32_t frequency, uint32_t duration)
{
  unsigned long toggle_count = 2*frequency*duration/1000000 - 8;
  tone(pin,frequency);
  timer2_toggle_count = toggle_count;
}

void ir_send_bit(uint8_t pin, uint16_t frequency, uint16_t
                                              on_time, uint16_t off_time)
```

```
{
  ir_send_pulse(pin,frequency,on_time);
  // Wait for the end of the pulse train.
  while (OCR2A!=0);
  // About 104 us are needed to compensate for the delay between the
  // call of ir_send_pulse and the beginning of the pulse train.
  delayMicroseconds(off_time-104);
}

void ir_send_code(uint8_t pin, uint16_t frequency, uint32_t code)
{
  ir_send_bit(pin,frequency,startbit,startbit/2);
  for (uint32_t mask=0x80000000; mask!=0; mask>>=1)
  {
    if ((code&mask)==0)
    {
      ir_send_bit(pin,frequency,on_time_0,off_time_0);
    }
    else
    {
      ir_send_bit(pin,frequency,on_time_1,off_time_1);
    }
  }
  ir_send_bit(pin,frequency,on_time_1,off_time_1);
}

void setup(void)
{
  bitClear(TIMSK0,TOIE0);
  ir_send_code(ir_led,CARRIER,RC_CODE);
  bitSet(TIMSK0,TOIE0);
}

void loop(void)
{
}
```

I opted for sending the code only once, in the function setup, so the function `loop` remained empty. Just press the reset button to resend the code. The function `ir_send_code` is called in `setup`. This call is between calls to the function `bitClear` and `bitSet` that are part of the Arduino API and that allow you to set or clear bits in a variable. Here they are used to temporarily halt timer/counter 0 while the code is being transmitted. As explained earlier in this chapter, timer/counter 0 has a period of about 1 ms, which is the same order of magnitude as our pulse train durations and it is possible to observe interference when viewing the output signal on an oscilloscope. Stopping timer/counter 0 removes this interference. Remember that the function `delayMicroseconds` used in `ir_send_pulse` does not rely on timer/counter 0, so stopping the latter is not really a problem.

The function `ir_send_pulse` begins with sending the start bit. Then it sends the data bits, the MSB first. The **`for`** loop here does not use a normal counter, but a 32-bit shift register (see also Section 8.5.1, page 198). The register is initialized with the value 0x8000 0000, which means that its most significant bit is set to one, while all the other bits are zero. During each loop iteration the value is shifted one position to the right and the new most significant bit is set to zero. The loop ends when the shift register contains only zeros, which is the case after exactly 32 iterations. The value of the shift register is also used to extract – through an AND operation – one by one the bit values of the code to be transmitted. Since this sketch sends a NEC-1 code, the function ends by sending the stop bit.

To send a bit, I wrote the `ir_send_bit` function that takes the mark and space lengths as arguments. The function `ir_send_bit` starts by calling the function `ir_send_pulse` to send the pulse, then waits for the pulse train to end before terminating by waiting the space length.

9.15.2 Exposure Time

Detecting the end of the pulse train is not that simple and requires another visit to the file `Tone.cpp`. When the function `ISR` has exhausted its level change budget, it calls `noTone` which in turn calls `disableTimer`. Since the latter clears the value of the OCR2A register, we can detect the end of the pulse train by reading this register regularly.

A second hack is necessary to ensure that the space will have the right duration. It is important here to take into account the delay between the call to the function `ir_send_pulse` and the beginning of the pulse train, which is more than 100 microseconds. This delay, if not handled properly, causes the *previous* space to be too long. To correct this, a value of 104 is subtracted from the space length in the call to the function `delayMicroseconds`. I measured the value of 104 with an oscilloscope but if you want to be more accurate, you can count the number of instruction cycles executed by the MCU between the two moments mentioned above.

The function `ir_send_pulse` also contains a subtlety. It starts with the calculation of the value for the variable `toggle_count`. This is done on purpose, because doing the calculation requires some time. Calling the function `tone` will make the pulse train depart. If we do the calculation after the departure of the train, it will continue to advance during the calculation. At 38 kHz a pulse train with a duration of 560 μs contains only 21 pulses, so if we want to change the value of `timer2_toggle_count` in time after having called `tone`, we better not waste too much time. Therefore, we do the computing first and then call `tone`.

The calculation contains a factor of eight that is subtracted from the result. This correction compensates for the time required to load the result of the calculation in the variable `timer2_toggle_count`. Since this variable is of type **`unsigned long`** that occupies four bytes, the MCU needs several instructions to copy all four, which takes time.

9.15.3 Capturing Volatile Moments

The variable `timer2_toggle_count` is declared as **`extern`** because it does not "live" in our sketch. I copied its definition (the part following the **`extern`** keyword) from the file `Tone.cpp`. The C keyword **`volatile`** that is part of this definition tells the compiler that the value of this variable can change at any moment meaning that the program must reload it every time the value is needed. The compiler may not expect that the value of such a variable will remain unchanged between the moment that it is loaded, at the beginning of a function for instance, and the moment that it is used. The register OCR2A too is declared as **`volatile`**.[1] If it hadn't be, the loop **`while`** `(OCR2A!=0)` would probably not have worked as intended. The reason for this is a program execution speed optimization performed by the compiler. If OCR2A was like any other variable, the compiler would make the MCU load a copy of it into one of its working registers before entering the **`while`** loop. In the loop, the MCU would use its copy instead of the real variable because that would save some instruction cycles. However, if the real variable is changed by an external event – a counter that overflows, for example – while the MCU is executing the loop, it will never see this change and it will continue looping around for the rest of its days. Specifying the keyword **`volatile`** forces the program to read a variable or register just before its contents are needed.

Compiler optimizations can be so effective that sometimes instructions may be eliminated altogether. Keep this in mind when you try for instance to slow down a piece of code by adding instructions that do not do anything. The compiler detects this and removes them, and your program will not be slowed down at all. Compilers are very powerful, sometimes even too powerful.

1. In fact, all the MCU registers are declared as **`volatile`**, because their values can be changed at any time by external and internal events like the arrival of a pulse or a data byte.

A final remark to close this section: if you look at the pulse train on an oscilloscope you will see that despite our efforts, it may not always contain the same number of pulses. This does not interfere with the proper functioning of the sketch, but it is annoying for us sticklers. This interference is due to the function `tone` that does not completely initialize timer/counter 2 because it does not reset the count register TCNT2. So, the time needed to reach the first level toggle is undefined and therefore the moment that we change the value of the toggle counter is undefined too. The result is that every once in a while the pulse train has a pulse too many. If you want to correct this weakness, add, in `tone`, the line `TCNT2 = 0` just before the line `OCR2A = ocr`.

9.16 Occupation: Rioter

You have probably already been in a similar situation: as part of a sports event, the local favorite team must play a decisive match, and you were invited by your friends to come over and watch the game at their place. Little by little, under the influence of alcoholic drinks and the exciting atmosphere in general, your friends start making more and more noise. Every time the favorite team attacks, they yell louder until, after a while, your ears begin to ache. Another scenario occurs when the favorite team is not playing as good as they should, is even outperformed by the opponent and your friends have become very quiet. Until suddenly, against all hope, the favorite team manages to launch a counter attack that looks very promising and may possibly limit the damage. Now your friends start screaming and shouting to support their team and, in doing so, make your ears ache. So how can you prevent this from happening again?

Declining the invitation would of course be unpolite, but fortunately you studied programming microcontrollers and you have developed a system that changes the TV channel when the ambient noise level becomes too high. Whenever your friends start screaming, the television will switch to another channel and they will not be able to see the outcome of the spectacular counter attack or, for that matter, any of the goals, dunks or home runs. As soon as your friends have figured out that it is you who is at the origin of this spoiled evening, they will stop inviting you over, allowing you to spend all your free time again on programing microcontrollers. So what do you need to build such a system?

First of all you will need a sound detector to measure the ambient noise level. Because you have a microcontroller at your disposition, all you need is a microphone with a preamplifier so that the MCU can digitize the microphone's output signal. The circuit diagram below shows what I came up with.

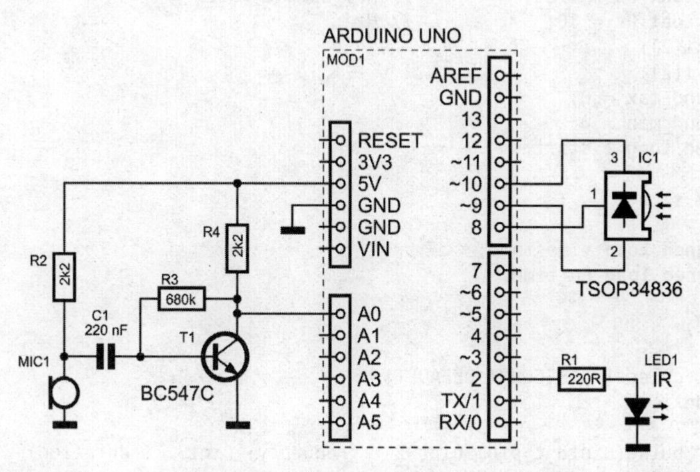

Figure 9-15 - The sound detector comprises an MCU and a single-transistor amplifier. The infrared receiver is the same as before and the infrared output is identical to the one in the previous example.

The microphone is an electret model that requires a little bit of power to work. The amplifier is just a transistor which in this configuration offers a gain of 100 (but only if the transistor is a C-type BC547 or another NPN transistor with an equivalent h_{fe}). The rest of the sound detector can be implemented in software, which is what we will do. The sketch must be able to record a code transmitted by an infrared remote control and replicate it on its infrared output when the ambient noise level exceeds a certain level. We have already written most of the program, only the sound detector remains to be implemented. Here is the sketch:

```
/*
 * sound detector
 */

#define CARRIER  37000
#define THRESHOLD  200
#define TIMEOUT_DEFAULT  2000000
#define PULSECOUNT_DEFAULT  50

int ir_signal = 8;
int ir_vcc = 10;
int ir_gnd = 9;
int mic = A0;
int led = 13;
int ir_led = 2;
extern volatile long timer2_toggle_count;
```

```
const float Fs = 1760;          // Hz, sample rate
const float Fc = 10;            // Hz
float ki[2];
float yi[2];
int sound_max = 0;
int sound_min = 0;
unsigned long t_start;

typedef struct
{
  unsigned long timestamp;
  unsigned long on_time;
}
point_t;

point_t pulses[PULSECOUNT_DEFAULT];
int index;

void ir_pulse(uint8_t pin, uint32_t frequency, uint32_t duration)
{
  unsigned long toggle_count = 2*frequency*duration/1000000 - 8;
  tone(pin,frequency,duration);
  timer2_toggle_count = toggle_count;
}

void ir_send_pulse(uint8_t pin, uint16_t frequency, uint16_t
                                              on_time, uint16_t off_time)
{
  ir_pulse(pin,frequency,on_time);
  while (OCR2A!=0);
  delayMicroseconds(off_time-104);
}

void ir_send(uint8_t pin, uint16_t frequency)
{
  unsigned long off_time = 1000;
  int i;
  bitClear(TIMSK0,TOIE0);
  for (i=0; i<index-1; i++)
  {
    off_time = pulses[i+1].timestamp - pulses[i].timestamp
                                              - pulses[i+1].on_time;
    ir_send_pulse(pin,frequency,pulses[i].on_time,off_time);
  }
  ir_send_pulse(pin,frequency,pulses[i].on_time,off_time);
  bitSet(TIMSK0, TOIE0);
}

int ir_receive(void)
{
  index = 0;
```

```
  unsigned long time_start = 0;
  unsigned long pulse = pulseIn(ir_signal,LOW,TIMEOUT_DEFAULT);
  while (pulse!=0 && index<=PULSECOUNT_DEFAULT)
  {
    unsigned long time_now = micros();
    if (time_start==0) time_start = time_now;
    pulses[index].timestamp = time_now - time_start;
    pulses[index].on_time = pulse;
    index += 1;
    pulse = pulseIn(ir_signal,LOW,TIMEOUT_DEFAULT);
  }
  return index;
}

void lpf_init(float fs, float fc, float coeffs[], float y[])
{
  coeffs[0] = exp(-TWO_PI*fc/fs);
  coeffs[1] = 1.0 - coeffs[0];
  y[0] = 0.0;
  y[1] = 0.0;
}

float lpf(int sample, float coeffs[], float y[])
{
  y[0] = coeffs[1]*sample + coeffs[0]*y[1];
  y[1] = y[0];
  return y[0];
}

boolean sound_detected(int threshold)
{
  int sound = analogRead(mic);
  sound += analogRead(mic);
  sound += analogRead(mic);
  sound += analogRead(mic);
  sound >>= 2;

  if (sound>sound_max) sound_max = sound;
  if (sound<sound_min) sound_min = sound;
  sound_max--;
  sound_min++;
  lpf(sound_max-sound_min,ki,yi);
  if (yi[0]>threshold) return true;
  return false;
}

void setup(void)
{
  pinMode(led,OUTPUT);
  digitalWrite(led,LOW);
  lpf_init(Fs,Fc,ki,yi);
```

```
  pinMode(ir_signal,INPUT);
  pinMode(ir_gnd,OUTPUT);
  digitalWrite(ir_gnd,LOW);
  pinMode(ir_vcc,OUTPUT);
  digitalWrite(ir_vcc,HIGH);
  // Suppress spurious pulses on power up.
  delay(500);
  while (pulseIn(ir_signal,LOW,500000)!=0);

  // Signal that receive window is open.
  digitalWrite(led,HIGH);
  ir_receive();
  digitalWrite(led,LOW);

  t_start = millis();
}

void loop(void)
{
  unsigned long t = millis();
  if (sound_detected(THRESHOLD)==true && t-t_start>5000)
  {
    digitalWrite(led,HIGH);
    ir_send(ir_led,CARRIER);
    digitalWrite(led,LOW);
    t_start = millis();
  }
}
```

In the function `loop` we wait for the ambient sound level to pass the threshold. When this happens, and if this did not happen in the previous five seconds, the recorded remote control code is transmitted. A flash of the LED on the Arduino board offers visual feedback. The choice of the period of five seconds is a bit arbitrary; the goal is to avoid that the system is triggered too often. In a real-life situation thirty seconds or one minute will probably be better, but for development purposes five seconds is already long enough.

The function `setup` is a kind of best-of compilation of the `setup` functions from the previous examples. It contains the configuration of the IR LED output and of the IR receiver with the removal of spurious impulses. Before handing over the execution of the sketch to the function `loop`, it allows the user to record a remote control code. The period during which this is possible is indicated by the on-board LED. In the function `setup` we also find the initialization of a low-pass filter similar to the one we used in Chapter 7 for the PID controller. In this sketch, the filter is used by the sound detector.

The function sound_detected uses the same method to set the sampling rate to almost 2 kHz as the PID controller sketch in Chapter 7. The brouhaha that will trigger the transmission of the RC code contains many low-frequency components and we do not need to sample at a high frequency.

At rest, the voltage measured at the MCU's input will be around 2.5 V, but its exact value depends on the amplifier. If we base our decisions on the amplitude of the signal instead of the difference between the average value and the actual value, we do not need to know the idle level. An extreme value detector (sound_min and sound_max) will do. The amplitude or the peak-to-peak value is filtered by the low-pass filter (the same as in Chapter 7) which has a relatively low cut-off frequency of 10 Hz. To ensure that the measured amplitude returns to its idle value, the extreme values sound_min and sound_max are not "impermeable". Indeed, whenever the function sound_detected is called, these two variables lose some of their values due to a small leak. The result is that the extreme values follow the ambient noise level, but with a little delay, and the measured amplitude will return to almost zero. Thus, a loud sound produces a good signal with a rising edge steeper than the trailing edge, a kind of asymmetric bell, which makes it easy to define a threshold.

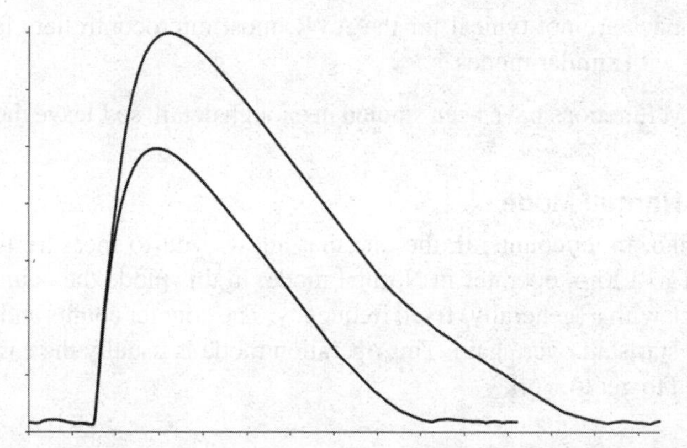

Figure 9-16 - The evolution of the amplitudes of two loud sounds picked up by the MCU.

The ir_receive function records up to PULSECOUNT_DEFAULT pulses. If this number is not reached in time, the function is automatically terminated thanks to the timeout argument of pulseIn. The pulses are stored in a table as before.

Reproducing the recorded code is simple, it suffices to reproduce the marks and spaces stored in the table. This is taken care of by the function ir_send in the same way as in the previous example.

9.17 Summarizing

After this long monograph on atomic clock receivers, infrared remote controls and how to imitate them all, you may have forgotten the theme of this chapter: using time in microcontroller applications. But, believe it or not, several important methods of using a timer/counter module have been presented, even though in a casual way. We did pulse width modulation (PWM), we measured pulse durations and event arrival times. Furthermore, we generated pulses with precisely defined lengths at specific instants in time; we generated pulse trains and signals with arbitrary frequencies. Actually we used most of the functions offered by the MCU's timer/counter modules and, what is more, we used the three modules available in the MCU of the Uno. So without you really noticing it, you learned a lot about counters and timers. Let's review them quickly.

The timer/counters modules of the AVR offer a total of four different modes:

- Normal;
- CTC;
- PWM;
- Capture.

These modes are not typical for the AVR, most microcontrollers feature similar modules with similar modes.

The PWM functions have been studied in enough detail, so I leave them aside here.

9.17.1 Normal Mode

In Arduino, timer/counter 0, the one that allows you to measure time using the function `millis`, operates in Normal mode. In this mode the counter is clocked by a clock with a (generally) fixed frequency. The counter counts and when it overflows, it starts all over again. This operation mode is usually the easiest to understand and to get to work.

9.17.2 CTC Mode

The Clear Timer on Compare match mode or CTC mode is almost identical to normal mode, except for the maximum value of the counter. In normal mode, this value is 255 for an 8-bit counter and 65,535 for a 16-bit counter. In CTC mode, on the other hand, the maximum value is specified by the OCR register. The function `tone` uses timer/counter 2 in this mode.

AVR microcontroller or not, 8-bit, 16-bit or 32-bit MCU, the way to use a timer/counter in normal or CTC mode is almost always the same. In one or more control registers you need to specify the source of the counter's clock and its frequency (or prescaler) and you may have to enable it. You also have to specify what the MCU must do when the counter's end condition is reached. Several options are possible like setting, clearing or toggling the level of a pin, clearing a register or generating an interrupt request. In CTC or similar mode you must of course specify the counter's end value.

9.17.3 Capture Mode

Capture mode allows you to automatically capture an event or a pulse. To do this, you should arm an input and specify what type of event – a rising or falling edge or a level – will trigger the capture hardware. When the event occurs, the counter value is copied into another register where you can retrieve it when you need it. Only the timer/counter on the ATmega328 features the capture function (the MCU on the Mega board has four of these modules), its input is pin 8 of the Uno board. The analog comparator which is not accessible from the API provides another way to start an automatic capture.

Curiously, the function `pulseIn` of the Arduino API does not use the MCU's capture functions, it is implemented entirely in software. The advantage is that this function can use any pin of the MCU as input. Also it does not consume any timer/counter resources because it counts instruction cycles (in fact, it counts the number of iterations of a loop with a known execution time). However, its disadvantage is that it blocks. The MCU cannot do anything else (except for responding to interrupt requests) while the function `pulseIn` waits for the event to happen. This particularity makes that outside of Arduino you will only rarely see this technique to measure time.

9.18 May The Force Be With You

Thanks to their flexibility and the many functions they offer, timer/counter peripherals and their registers are often a bit difficult to penetrate but, thanks to the things you learned in this chapter, you should be able to understand how to use them. Or, as Saint Augustine put it, "*Let us see then, thou soul of man, whether present time can be long: for to thee it is given to feel and to measure length of time.*"

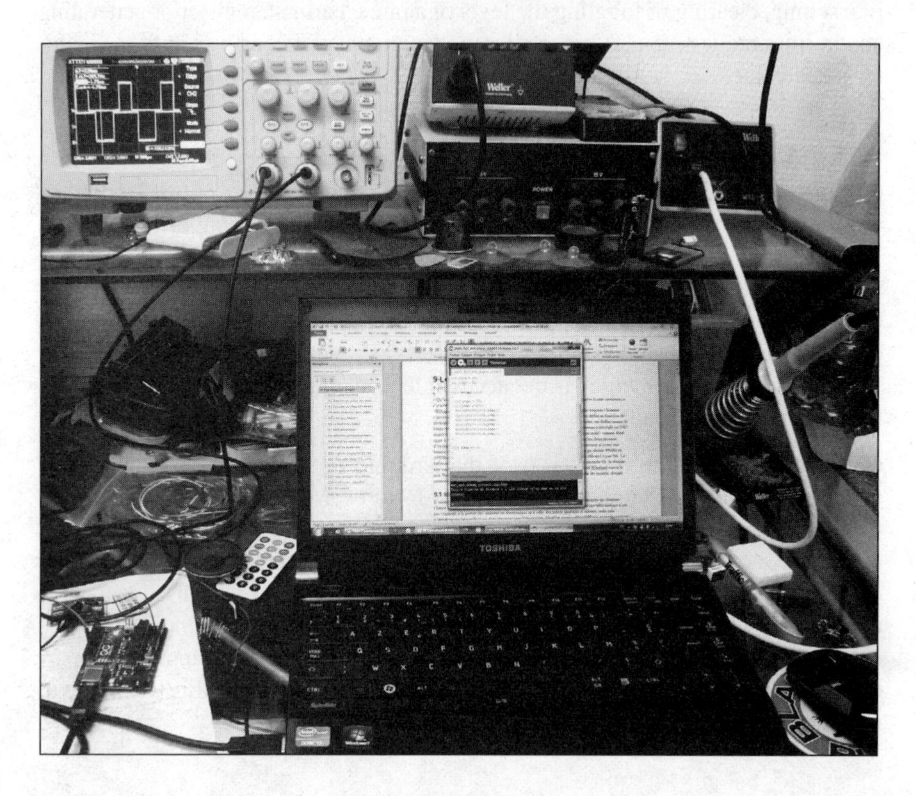

The author's desktop.

10. Interrupts - Pandora's Box

So far the programs presented in this book were pretty deterministic, meaning that at any time, if we knew the states and values of all variables, we knew exactly how the program would behave. When a program needed some sort of a signal to launch an operation, it waited for the arrival of the signal by scanning an input, a variable or a bit. However, while the program was waiting, the MCU could not really do anything else but twiddle its thumbs. That this is not very efficient in general is one thing, but more important is the fact that in many situations it is not a good idea to block the MCU to wait sheepishly for an event to happen. Consider a system where a microcontroller should respond differently to two (or more) events that can occur at any time. If event A always happens before event B, it is possible to wait first for A and then wait for B. However, when events are independent and event A occurs while the MCU is stuck in a loop waiting for event B, it cannot respond to event A, the LED will not be switched on or, worse, the cooling system is not started, the nuclear power plant overheats and the new episode of your favorite television series is interrupted by an alert from the Government asking you to remain calm and close doors and windows to prevent an unreasonable daily intake of nuclear radiation. Did you know that nuclear radiation has so-called deterministic effects on living organisms? Determinism decidedly has its drawbacks.

Interrupts were invented to allow microcontroller system designers to escape from delicate scenarios like the one described above. An interrupt is a signal that is meant to make the MCU respond to an event. The MCU will interrupt its current activity and execute a special function instead. The MCU is supposed to drop everything to respond as quickly as possible to interrupts. This is just like you and your phone; you will drop everything so that you can answer your phone as quickly as possible. The MCU does the same for interrupts. And, like you, it does not let its cup of coffee fall out of its hands when an interrupt occurs. No, it puts it on the table first, meaning that it needs a (very) short delay before it can respond to the interrupt. Fortunately, putting down its coffee does not take a lot of time and the MCU is usually able to react within a few microseconds. When the MCU has finished processing the interrupt, it returns to its current affairs.

Many novice programmers are afraid of interrupts, yet they are not too difficult to master if certain rules are respected. For proper interrupt management it is useful – and even essential – to understand how they work.

10.1 My First Interrupt

The MCU has a table which lists for each known or authorised interrupt the address – the vector – of the function to execute when the associated interrupt occurs. When this happens, the MCU terminates the execution of the instruction at hand; it notes in a special memory what it was doing (its state, i.e. the values of some important registers) at the time of the interrupt, looks in the table to find out where it is supposed to go to respond to the interrupt in question, quickly jumps to this location and starts executing the code that lives there. The function to be executed, the Interrupt Service Routine or ISR[1], is supposed to be finite so that the MCU can resume its activities it was busy with before the interrupt occurred, but this is actually not mandatory.

10.1.1 Timer/Counter 0

Here is a simple sketch that uses interrupts:

```
/*
 * interrupts 1
 */

void setup(void)
{
}

void loop(void)
{
}
```

No, I'm not kidding you. On the contrary! Even though nothing seems to happen in this sketch, its execution is interrupted almost every millisecond by timer/counter 0. As I explained in the previous chapter, in Arduino this counter is reserved for the function `millis` and its friends. The counter is started when a sketch is launched and, without user intervention, it will only stop when you turn off the power to the MCU.

In Arduino this 8-bit counter is clocked by a signal of 250 kHz (16 MHz / 64) and it counts from 0 to 255 (corresponding to 256 values). When the counter overflows, it produces an interrupt request which causes the microcontroller to execute the associated interrupt service routine. This function, with the romantic name

1. Interrupts are like guests and clients, they are served immediately and always with a smile.

SIGNAL(TIMER0_OVF_vect) and that can be found in the file wiring.c, is therefore called almost every millisecond to update the millisecond counter. Since it is called so often, it must contain as few statements as possible to terminate as quickly as possible. The longer its execution time, the bigger its impact on the overall execution speed of the program. This brings us to an important rule for interrupt service routines: always make sure that their execution time is as short as possible. Since this rule is so important, I will repeat it in italic:

✏ *The execution time of an interrupt service routine must be as short as possible.*

10.1.2 Generating a 1 kHz Signal

Respecting this important rule is not always easy. Take for example the case of Arduino. The interrupt service routine for timer/counter 0 is relatively long as it must correct the millisecond counter for its clock frequency of 976.6 Hz that is slightly too low. This complication is due to the desire to use this counter for two things at once: counting milliseconds and generate PWM signals. CTC mode (see Section 9.17.2, page 282) allows an interrupt frequency of exactly 1 kHz (16 MHz / 64 / 250), but only by curbing the PWM part. If you do not need the PWM capabilities of the pins 5 and 6 (on the Uno, pins 4 and 13 on the Mega) and if the lack of precision of millis bothers you, you can change the behavior of the counter, like so:

```
/*
 * interrupt 2
 */

int timer0 = 6;
int led = 13;
unsigned long time_last;

extern volatile unsigned long timer0_overflow_count;
extern volatile unsigned long timer0_millis;

SIGNAL(TIMER0_COMPA_vect)
{
  timer0_millis++;
  timer0_overflow_count++;
}

void led_toggle(void)
{
  static int led_value = 0;
  digitalWrite(led,led_value);
```

```
  led_value ^= 1;
}

void setup(void)
{
  pinMode(timer0,OUTPUT);
  pinMode(led,OUTPUT);
  bitClear(TCCR0A,WGM00);
  bitSet(TCCR0A,COM0A0);
  OCR0A = 249;
  bitSet(TIMSK0,OCIE0A);
  time_last = millis();
}

void loop(void)
{
  unsigned long time = millis();
  if (time-time_last>=10)
  {
    led_toggle();
    time_last = time;
  }
}
```

The function `loop` only flashes the on-board LED at a frequency of 100 Hz to illustrate the improved accuracy of `millis` (check with a frequency counter or an oscilloscope). To turn on and off the LED with a single function, I created the function `led_toggle`. This function has nothing special except for the variable `led_value`. I declared it as **`static`**, which means that it retains its value between function calls. It is like declaring it as a global variable, as in this sketch the variable `timer0`, except that `led_value` is not accessible outside the function `led_toggle`. The C keyword **`static`** allows you to create a mutant variable that is semi-local, semi-global.

For once the interesting part of this sketch is in the function `setup` and also partly in the interrupt service routine `SIGNAL`. Even though it looks like this, the function `setup` is not the beginning of the sketch and the MCU has been initialized earlier in the function `init` contained in the file `wiring.c`. By studying this function we discover that counter 0 is activated in mode 3, fast PWM. We need mode 2, CTC. To achieve this, it is enough to clear the WGM00 bit of the TCCR0A register, for example by using the API function `bitClear`. And now that we are changing the register's contents, we might as well set its COM0A0 bit to activate a control signal on pin 6 that we configured at the top of `setup` as an output. This signal will allow us to check the frequency of counter 0 with an oscilloscope. For an interrupt frequency of 1 kHz we must divide the 250 kHz clock signal of counter 0 by 250. Since the counter always starts at zero, in order to divide its frequency by 250, the counter must count up to 249. This is the value that must be written in the OCR0A comparison register. On the oscilloscope you will observe a frequency of 500 Hz, because the control signal changes level on each successful comparison which effectively halves the signal's frequency.

Now that we are using CTC mode the overflow interrupt is no longer occurring because the counter does not overflow anymore, which prevents the interrupt service routine `SIGNAL(TIMER0_OVF_vect)` from being called. Instead, we activate the comparison interrupt flag OCIE0A that is raised when the counter 0 reaches the value stored in the OCR0A register. The corresponding interrupt service routine is `SIGNAL(TIMER0_COMPA_vect)` and this is where we now count the milliseconds needed by the function `millis`. We must also count overflows because this value is used by the function `micros`. Both variables `timer0_millis` and `timer0_overflow_count` are declared as **`extern`** since they belong to the file `wiring.c` that does not expose them, i.e. their definitions are not published to make them accessible to other files.

Note that the function `micros` does not work quite correctly anymore in this sketch. It is supposed to calculate from the counter overflows the number of microseconds that have passed since the beginning of the sketch. However, `micros` assumes that this counter is updated every 1024 µs (976.6 Hz) instead of every 1000 µs (1 kHz) and thus its calculation is wrong. You will have to divide the result by 1024 / 1000 = 1.024 to obtain the correct value.

10.2 The Devil in Disguise

Now look more closely at the name of the interrupt service routine `SIGNAL`. In fact, it is neither a function name nor a reserved word of C, but a macro that hides the directives that allow the compiler to understand that this function is not just any function, but an interrupt service routine. This distinction is necessary because microcontrollers often have a special instruction to return from an interrupt service routine. Furthermore, these directives specify whether the compiler must add additional instructions to save certain registers at function entry and restore them at exit. Since an interrupt can happen at any time and the work of the MCU can be interrupted at any stage, in the middle of a calculation for example, it is essential to store the values of some registers before executing an interrupt service routine that potentially may overwrite the contents of these registers. Unfortunately, saving register values takes a bit of time. In some systems it is essential that the MCU responds as quickly as possible to interrupts and every microsecond that can be spared is important. By carefully writing the interrupt service routine it is possible to limit the number of registers it uses, which also limits the number of registers to be saved before executing the routine. However, doing so can be very risky and it has crashed many a program, therefore I will not tell you how to do it. In the world of Arduino the standard response time to interrupts is usually more than sufficient.

In the previous chapter, I briefly mentioned a function called `ISR`. In fact, as I can reveal now, `ISR` is not a function name, it is not a C keyword either, but it is a macro like `SIGNAL`. More exactly, `ISR` is similar to `SIGNAL`, but more powerful. It is advisable to use only `ISR` and forget `SIGNAL`, precisely because the latter is less powerful. `ISR` stands for Interrupt Service Routine, but you had probably guessed that already, hadn't you?

10.2.1 What's Our Vector, Victor?

The argument of `ISR` (and `SIGNAL`) is the vector – or address, if you prefer – of the interrupt service routine. The vector must obviously be associated to the interrupt signal, otherwise nothing will happen. The MCU on the Uno features 25 of these vectors; the one on the Mega board has 56 of them.

These two tables show that each peripheral integrated in the microcontroller can generate at least one interrupt request.

Of all these vectors only `TIMER0_OVF_vect` is never available in Arduino as it is used by the API. When you use the serial port in your sketch, the vectors `USART0_RX_vect` and `USART0_UDRE_vect` will also become unavailable. If you try to use an interrupt vector that is already in use by a library or an API function, the compiler – or rather the linker – will report a "multiple definition" error.

No.	Vector	Description
1	INT0_vect	External Interrupt Request 0
2	INT1_vect	External Interrupt Request 1
3	PCINT0_vect	Pin Change Interrupt Request 0
4	PCINT1_vect	Pin Change Interrupt Request 1
5	PCINT2_vect	Pin Change Interrupt Request 2
6	WDT_vect	Watchdog Time-out Interrupt
7	TIMER2_COMPA_vect	Timer/Counter2 Compare Match A
8	TIMER2_COMPB_vect	Timer/Counter2 Compare Match B
9	TIMER2_OVF_vect	Timer/Counter2 Overflow
10	TIMER1_CAPT_vect	Timer/Counter1 Capture Event
11	TIMER1_COMPA_vect	Timer/Counter1 Compare Match A
12	TIMER1_COMPB_vect	Timer/Counter1 Compare Match B
13	TIMER1_OVF_vect	Timer/Counter1 Overflow
14	TIMER0_COMPA_vect	Timer/Counter0 Compare Match A
15	TIMER0_COMPB_vect	Timer/Counter0 Compare Match B
16	TIMER0_OVF_vect	Timer/Couner0 Overflow
17	SPI_STC_vect	SPI Serial Transfer Complete
18	USART_RX_vect	USART Rx Complete
19	USART_UDRE_vect	USART Data Register Empty
20	USART_TX_vect	USART Tx Complete
21	ADC_vect	ADC Conversion Complete
22	EE_READY_vect	EEPROM Ready
23	ANALOG_COMP_vect	Analog Comparator
24	TWI_vect	I²C / Two-wire Serial Interface
25	SPM_READY_vect	Store Program Memory Read

Table 10-1 - The interrupt vectors of the microcontroller that equips the Uno board.

No.	Vector	Description
1	INT0_vect	External Interrupt Request 0
2	INT1_vect	External Interrupt Request 1
3	INT2_vect	External Interrupt Request 2
4	INT3_vect	External Interrupt Request 3
5	INT4_vect	External Interrupt Request 4
6	INT5_vect	External Interrupt Request 5
7	INT6_vect	External Interrupt Request 6
8	INT7_vect	External Interrupt Request 7
9	PCINT0_vect	Pin Change Interrupt Request 0
10	PCINT1_vect	Pin Change Interrupt Request 1
11	PCINT2_vect	Pin Change Interrupt Request 2
12	WDT_vect	Watchdog Time-out Interrupt
13	TIMER2_COMPA_vect	Timer/Counter2 Compare Match A
14	TIMER2_COMPB_vect	Timer/Counter2 Compare Match B
15	TIMER2_OVF_vect	Timer/Counter2 Overflow
16	TIMER1_CAPT_vect	Timer/Counter1 Capture Event
17	TIMER1_COMPA_vect	Timer/Counter1 Compare Match A
18	TIMER1_COMPB_vect	Timer/Counter1 Compare Match B
19	TIMER1_COMPC_vect	Timer/Counter1 Compare Match C
20	TIMER1_OVF_vect	Timer/Counter1 Overflow
21	TIMER0_COMPA_vect	Timer/Counter0 Compare Match A
22	TIMER0_COMPB_vect	Timer/Counter0 Compare Match B
23	TIMER0_OVF_vect	Timer/Counter0 Overflow
24	SPI_STC_vect	SPI Serial Transfer Complete
25	USART0_RX_vect	USART0, Rx Complete
26	USART0_UDRE_vect	USART0 Data register Empty
27	USART0_TX_vect	USART0, Tx Complete
28	ANALOG_COMP_vect	Analog Comparator

Table 10-2 - The interrupt vectors available in the MCU of an Arduino Mega board.

No.	Vector	Description
29	ADC_vect	ADC Conversion Complete
30	EE_READY_vect	EEPROM Ready
31	TIMER3_CAPT_vect	Timer/Counter3 Capture Event
32	TIMER3_COMPA_vect	Timer/Counter3 Compare Match A
33	TIMER3_COMPB_vect	Timer/Counter3 Compare Match B
34	TIMER3_COMPC_vect	Timer/Counter3 Compare Match C
35	TIMER3_OVF_vect	Timer/Counter3 Overflow
36	USART1_RX_vect	USART1, Rx Complete
37	USART1_UDRE_vect	USART1 Data register Empty
38	USART1_TX_vect	USART1, Tx Complete
39	TWI_vect	I²C / Two-wire Serial Interface
40	SPM_READY_vect	Store Program Memory Read
41	TIMER4_CAPT_vect	Timer/Counter4 Capture Event
42	TIMER4_COMPA_vect	Timer/Counter4 Compare Match A
43	TIMER4_COMPB_vect	Timer/Counter4 Compare Match B
44	TIMER4_COMPC_vect	Timer/Counter4 Compare Match C
45	TIMER4_OVF_vect	Timer/Counter4 Overflow
46	TIMER5_CAPT_vect	Timer/Counter5 Capture Event
47	TIMER5_COMPA_vect	Timer/Counter5 Compare Match A
48	TIMER5_COMPB_vect	Timer/Counter5 Compare Match B
49	TIMER5_COMPC_vect	Timer/Counter5 Compare Match C
50	TIMER5_OVF_vect	Timer/Counter5 Overflow
51	USART2_RX_vect	USART2, Rx Complete
52	USART2_UDRE_vect	USART2 Data register Empty
53	USART2_TX_vect	USART2, Tx Complete
54	USART3_RX_vect	USART3, Rx Complete
55	USART3_UDRE_vect	USART3 Data register Empty
56	USART3_TX_vect	USART3, Tx Complete

Table 10-2 - The interrupt vectors available in the MCU of an Arduino Mega board. (end)

10.3 Message in a Bottle

Timer/Counter 0 is not the only peripheral that uses interrupts without your prior written consent. The serial port also uses interrupts to perform certain tasks, but not in the same way as the counter does. The serial port uses an interrupt to automatically send its data without making the sketch wait, or, as seasoned programmers say, without blocking the program. It is like putting a letter in the mail: you write it, you drop it in the mailbox and you go on with your daily business. As you continue your life, the postal services will take care of transporting your letter to its destination. Apart from the case of recorded mail, the postal services will not bother you with any questions or progress reports and you are not informed of the delivery of the letter. In Arduino sending data over the serial port works pretty much in the same way. The sketch posts its data by calling for example the function `Serial.print` and the serial port does the rest without disturbing the sketch. So how does this work?

It is actually quite simple. When the sketch calls the function `Serial.print`, the data to be transmitted is copied into a buffer. Then the first data byte is loaded into the transmit register. Typically this causes the serial transmitter hardware to push out the data bits one at a time on the serial port's output[1]. Once all the bits have been sent, the serial port peripheral triggers an interrupt to indicate that its transmit register is empty. If there is more data to be sent, the interrupt service routine will load the next byte in the transmit register and the serial port peripheral will crunch out a new series of bits. This process continues without any intervention of the sketch until all data has been sent and the buffer is empty again.

Here is a sketch to show that I am not spinning yarns:

```
/*
 * interrupt 3
 */

int led = 13;

void setup(void)
{
  Serial.begin(1200);
  pinMode(led,OUTPUT);
  digitalWrite(led,HIGH);
  Serial.println("Six sick hicks nick six slick bricks with picks
                                                        and sticks");
```

1. The Arduino API starts a transmission by activating the transmit register empty interrupt.

```
}

void loop(void)
{
  digitalWrite(led,LOW);
}
```

In this sketch, the serial port runs at a low speed of 1200 baud, which means that only 120 characters can be sent every second (one start bit, eight data bits, no parity bit and a stop bit make a total of ten bits per character). The sentence has 58 + 2 characters (end of line and carriage return), meaning that 500 ms are needed to transmit them all. If the function `Serial.println` blocks the MCU, the LED should light up for almost half a second. If, on the other hand, `Serial.println` does not block, the LED should just flash.

As you have probably already guessed, when you execute this sketch the LED lights up only very briefly, thus confirming my (hypo)thesis. In addition, on my Uno board an LED is connected to the serial port signal TXD. Unlike the LED connected to pin 13, this LED does come on for half a second, which corroborates my theory.

Note that I could have put the only statement of the function `loop` at the end of the function `setup`, just after the `Serial.println` statement, but I think it is more convincing this way.

Reception of data also uses an interrupt, which allows you to automatically receive data. Each time a character is received by the serial port peripheral, it issues an interrupt request to make the associated interrupt service routine copy the character into a receive buffer. This allows the sketch to check the receive buffer – with `Serial.available` – every once in a while instead of continuously to see if data has been received.

10.4 Spinning Out Of Control

The interrupt service routines of timer/counter 0 and the serial port(s), and also the one used by the function `tone` (timer/counter 2 overflow interrupt), are inherent to Arduino and the API does not provide a method to use or modify them. The following four interrupt-related functions offered by the API do not allow this either:

- `interrupts`
- `noInterrupts`
- `attachInterrupt`
- `detachInterrupt`

The first two are used to enable or disable interrupt requests globally. All the microcontrollers and microprocessors that I have had the honor of meeting during my life offered the possibility to disable all interrupt requests with a single instruction[1]. Generally this is done by setting or clearing a bit in a configuration register of the MCU that controls the connection of the interrupt manager. When a microcontroller is turned on, interrupt requests are always disabled for reasons of security. Subsystems which are powered up at the same time as the MCU may produce spurious signals that should not be taken into account, while other signals – real interrupt requests – may not be treated before the system initialization procedure has been completed. Take for example the nuclear power plant at the beginning of this chapter. Imagine that during the power-up sequence of a nuclear power plant the fuel rod assembly interrupts the partly-awoken microcontroller to inform the latter that it is OK to immerse the rods in the cooling pond and that – as far as the fuel rod assembly is concerned – fission can begin, the MCU had better be sure that filling the cooling pond has indeed been completed before responding positively to the interrupt. If it did not do this check, you might again be forced to close doors and windows and have your favorite TV series interrupted by a health alert from the Government.

Calling the function `noInterrupts` disables all interrupt requests. This implies that if the sketch calls this function and then attempts to send data over the serial port, nothing will come out. In addition, the millisecond counter will no longer work, as will the function `delay` and all other functions that depend on timer/counter 0. Have a look at the following sketch. Since I disabled the interrupt requests in the function `setup` without reactivating them later, this sketch fails miserably. The LED does not flash and no characters appear at the output of the serial port.

```
/*
 * interrupt 4
 */

int led = 13;

void setup(void)
{
  Serial.begin(115200);
  pinMode(led,OUTPUT);
  noInterrupts();
  Serial.println("LED no blink no");
}
```

1. Except for the Non-Maskable Interrupt or NMI, if available.

```
void loop(void)
{
  digitalWrite(led,LOW);
  delay(250);
  digitalWrite(led,HIGH);
  delay(250);
}
```

The function `interrupts` (re)enables interrupt requests globally. This does not mean that calling this function will enable all possible interrupt requests; no, only the interrupt requests that have been activated explicitly at the peripheral level, will (again) be able to interrupt the microcontroller.

The pair `interrupts` - `noInterrupts` is especially useful when the MCU must not be interrupted at all in order to accurately respect a delay or timing sequence, or to avoid a peripheral interfering while the MCU performs a task. The first case is easy to understand. If you want to produce on one of the MCU's pins a pulse sequence with well-defined lengths and spaces, it is better to not allow the MCU being interrupted at random moments. When the MCU executes an interrupt service routine it cannot do anything else, making it difficult to respect delays.

Another reason to globally disable interrupt requests from time to time is when the MCU shares something with a peripheral such as a memory location, a register or a pin. To illustrate this, imagine the following scenario. The MCU wants to write a 32-bit integer to memory. Since on an 8-bit system this corresponds to writing four bytes, a sequence of several instructions has to be executed, and such sequences may be interrupted. Now imagine that at the same time a device, say a temperature monitoring system of a nuclear power plant, realizes that it needs this same 32-bit integer to, say, calculate the maximum temperature of the fuel rod cooling pond. The device interrupts the MCU that just finished updating the first byte of the 32-bit integer and has begun writing the second; the updates of the third and fourth bytes are still pending. As soon as the MCU has finished writing the second byte it will respond to the interrupt request. The MCU reads the four bytes from memory to fulfil the request of the temperature monitoring system and then returns to its office where it continues with the update of the third and fourth bytes of the 32-bit integer. The temperature monitoring system does not know that the data that it has received is in fact composed of two updated bytes and two stale bytes. The value that it calculates using this data will be wrong, the nuclear reactor will overheat and a security alert from the Government interrupts once more your favorite television series. That's three times; enough already!

Since interrupt requests may occur at any time, it is important to manage them carefully. They are convenient to perform tasks in the background, but, as their name suggests, they upset the course of the program. In short, interrupts have advantages and disadvantages. It is up to you, the programmer, to control and decide which peripheral will have the right to interrupt the MCU and under what circumstances. Interrupts form a formidable weapon in the programmer's arsenal, but be careful, because, as Spiderman so rightly said, with great power comes great responsibility.

10.5 Knock on Any Door

The MCU on the Uno board has two special inputs that can trigger an interrupt request. The MCU on the Mega board has eight of them. These interrupt requests, named INTx where x is a number from 0 to 7, can be triggered when the level on the pin changes, or has dropped low. In the latter case, interrupt requests are generated as long as the level remains low. In addition to these two (or eight) special inputs, 24 ports of the MCU (i.e. all ports of the Uno's MCU) [1] can also be used as external interrupt source, but only to detect level changes. These interrupts are called PCINTx, where x is a number from 0 to 23. PC stands for Pin Change, short for "when the level on a pin changes". PCINTx interrupts have the particularity that they work for both inputs and outputs, which means that you can trigger an interrupt request by changing the level of a pin that is configured as an output.

Curiously, the Arduino API only supports interrupts of the INTx type. They are easy to use thanks to the functions `attachInterrupt` and `detachInterrupt`. The first one takes as arguments the number x of the INTx interrupt, the name of the interrupt service routine and the type of event that can trigger the interrupt request. Four events are defined: `LOW` (a low level), `CHANGE` (a change of level), `RISING` (a rising edge) and `FALLING` (a falling egde). The interrupt service routine cannot take arguments and it cannot return a result. The function `detachInterrupt` only needs the interrupt number.

10.5.1 Let's Make a Flip-Flop

Here is a sketch that turns the MCU into a set-reset flip-flop. Pins 2 and 3 are respectively the set and reset inputs, pins 12 and 13 are the not-Q and Q outputs.

1. In fact, the Uno does not have PCINT15 because port PC7 has no physical connection, a pin, to the outside world. Also PCINT6 and PCINT7 are unavailable because PB6 and PB7 are used for the crystal. PCINT14 is available in theory, but it shares its port PC6 with the reset pushbutton. On the Mega board interrupts PCINT11 to 15 are unavailable.

```
/*
 * interrupt 5
 */

int Q = 13;
int not_Q = 12;
int S = 2;
int R = 3;

void reset(void)
{
  digitalWrite(Q,LOW);
  digitalWrite(not_Q,HIGH);
}

void set(void)
{
  digitalWrite(not_Q,LOW);
  digitalWrite(Q,HIGH);
}

void setup(void)
{
  pinMode(Q,OUTPUT);
  pinMode(not_Q,OUTPUT);
  digitalWrite(Q,LOW);
  digitalWrite(not_Q,HIGH);
  pinMode(S,INPUT_PULLUP);
  pinMode(R,INPUT_PULLUP);
  attachInterrupt(0,set,LOW);
  attachInterrupt(1,reset,LOW);
}

void loop(void)
{
}
```

As you can see, the function `loop` does nothing. The initialization of the flip-flop takes place in the function `setup`. The flip-flop itself is constructed from two interrupt service routines `set` and `reset`. The interrupt service routines are "installed" in the function `setup` using two calls of the function `attachInterrupt`.

If the inputs INT0 and INT1 are both low, i.e. active, the LED is switched on. This means that INT0 is served before INT1 or, in IT lingo, the priority of INT0 is higher than the priority of INT1. As a matter of fact, the position in the interrupt vector table determines the priority of an interrupt. The higher up in the table it is, the higher its priority. INT0 therefore has the highest priority of all.

Now let's change the reset interrupt service routine – not to be confounded with the MCU reset routine – by adding an infinite loop, like this:

```
void reset(void)
{
  digitalWrite(Q,LOW);
  digitalWrite(not_Q,HIGH);
  while (true);
}
```

Now fiddle a bit with the flip-flop's inputs. What do you notice? Indeed, as soon as INT1 is activated to switch the LED off, it is impossible to turn it back on again. How is this possible if INT0 has a higher priority than INT1? Well, simply because when the MCU responds to an interrupt request, the first thing it does is disabling all other interrupt requests so that it can process the current interrupt without being interrupted. Interrupt requests are re-enabled automatically at the end of the interrupt service routine. This is due to the MCU hardware that was designed to do this; it is not a special Arduino feature. By default, it is not possible to interrupt an interrupt service routine. To do it anyway, because it is possible to do it, you must manually and therefore explicitly enable interrupts in the interrupt service routine, as follows:

```
void reset(void)
{
  digitalWrite(Q,LOW);
  digitalWrite(not_Q,HIGH);
  interrupts();
  while (true);
}
```

10.6 One Interrupt Too Many

Because by default it is not possible to nest interrupts (interrupt nesting), the serial port is halted while the microcontroller executes an interrupt service routine, as is timer/counter 0 and the function tone. Therefore you should not forget to enable interrupts after entering an interrupt service routine if it is really necessary that everything continues all the time. On the other hand, if you do this, you will let yet another evil genius escape from Pandora's Box of Interrupts: the stack overflow.

Figure 10-1 - The optimistic programmer.

10.6.1 The Stack

The MCU has a special memory called the stack, which endorses more or less the role of mediator between the main program and the functions and interrupt service routines. This memory is called a stack because – in theory – you can only access it from the top. It is a kind of data stack where the last data written ("pushed" on the stack) is the first to be read ("popped" off the stack). The stack is used to pass arguments to functions, for storing local variables and also when the MCU has to leave its post to execute a function or an interrupt service routine. The stack also contains the address of the instruction that the MCU had planned to execute if only it had not been interrupted to execute this or that function or interrupt service routine first. The MCU is a real Tom Thumb, dropping addresses on the stack like white pebbles to allow it to retrace its way to where it was before it was called to do something else.

Like all the MCU's resources the stack also has its limits. Because the stack is often a part of the data memory, which is not very abundant to begin with, we tend to keep the size of the stack as small as possible. Consequently, if the program is running normally, the stack must have enough room to store the arguments, local variables and return addresses of all the functions of the longest call chain possible in the program. So, if a function calls another, which in turn calls a third and so on, the stack must be large enough to keep track of all the return addresses. Function arguments and local variables are also stored on the stack, which implies that functions with many and/or large arguments and/or local variables need a lot of stack space.

The stack overflows when there is not enough space to put new arguments, new local variables or a new return address. Depending on how the manufacturer designed the stack, an overflow may either result in overwriting memory that is not part of the stack, but that is located just behind the stack, or in overwriting data already on the stack. If the first case is not always fatal, the second usually is. The MCU will use a corrupted return address, continue executing instructions at a random position and show erratic behavior.

In a system with many interrupt sources of which some can repeat at high frequencies, the stack may quickly run out of space, especially if interrupts can be nested. All that is needed to overflow the stack is one interrupt request at the wrong time and there you are again shutting doors and windows because an alert from the Government suggested you to do so. So, in the end, the decision of the MCU manufacturer to automatically disable interrupt requests on entering an interrupt service routine was not that bad, was it?

10.7 Who's That Knocking At My Door?

Since the Arduino API does not support PCINTx interrupts we are forced to write our own functions to use them. We have already studied how to create a simple interrupt service routine, but we have not yet seen how to handle multiplexed interrupts. As the interrupt vector table shows, there are only three vectors for 24 PCINT interrupts. They are divided into three groups of eight, each with their own interrupt vector. Multiplexing interrupts is a common practice that reduces the number of vectors required, and often all the interrupt sources of the same peripheral are grouped in a single vector and one interrupt service routine. When an interrupt occurs, the interrupt service routine must first read the interrupt register of the peripheral in question to determine which interrupt it should process.

Unfortunately, for PCINT interrupts things are not so simple. The AVR does not have a register that shows which pin caused a PCINT interrupt, but we can create one ourselves. This requires storing the states of the inputs before an interrupt occurs so that they can be compared to the input states after an interrupt. Since each level change on an input triggers an interrupt, it is easy to identify which input is guilty. However, two difficulties present themselves here. First of all, there may be more than one input that changed level at the same time and you have to treat them all one by one. Secondly, not all inputs may be authorised to request interrupts. Handling the second difficulty is possible by using mask registers. PCINT inputs can be activated individually in three mask registers PCMSKx (Pin Change Mask) and every PCINT vector has such a mask. Before deciding whether or not an input is responsible for an interrupt, a logical AND between our homemade input state register and the associated mask register is required.

On the Uno board PCINT0 to 5 correspond to ports PB0 to PB5 with mask register PCMSK0. PCINT8 to 14 correspond to ports PC0 to PC6 (PJ on the Mega) and the mask register is PCMSK1. PCINT16 to 23 correspond to ports PD0 to PD7 (PK on the Mega) with mask register PCMSK2. The Uno does not offer access to PCINT6, PCINT7 and PCINT15. PCINT14 is available in theory, but shares its pin with the reset pushbutton. On the Mega PCINT11 to 15 are not accessible.

10.7.1 Multiplexed Interrupts

Here is a sketch that illustrates multiplexed PCINT interrupts:

```
/*
 * interrupt 6
 */

uint8_t pcint0_states_prev;
String result;

ISR(PCINT0_vect)
{
  uint8_t pin_states = PINB;
  uint8_t pcint = (pin_states ^ pcint0_states_prev) & PCMSK0;
  pcint0_states_prev = pin_states;
  if (pcint&0x01) result += '0';
  if (pcint&0x02) result += '1';
  if (pcint&0x04) result += '2';
  if (pcint&0x08) result += '3';
  if (pcint&0x10) result += '4';
  if (pcint&0x20) result += '5';
  result += '-';
}

void setup(void)
{
  Serial.begin(115200);
  pinMode(8,INPUT_PULLUP);
  pinMode(9,INPUT_PULLUP);
  pinMode(10,INPUT_PULLUP);
  pinMode(11,INPUT_PULLUP);
  pinMode(12,INPUT_PULLUP);
  pinMode(13,INPUT_PULLUP);
  pcint0_states_prev = PINB;
  PCMSK0 = 0x3f;                // Enable PCINT inputs.
  bitSet(PCIFR,PCIF0);          // Clear any pending interrupts.
  bitSet(PCICR,PCIE0);          // Enable PCINT0-7.
  result = "";
}
```

```
void loop(void)
{
  if (result.length()!=0)
  {
    Serial.println(result);
    result = "";
  }
}
```

The function `loop` checks each time the length of the string managed by the object `result`. This object acts as a data buffer for the characters "received" by the interrupt service routine of the sketch. As soon as `loop` detects the presence of one or more characters in the buffer, it copies them to the serial port and empties the buffer.

In the function `setup` all available pins of port B on the Uno board are configured as inputs with pull-up resistor and their current states are copied to the variable `pcint0_states_prev`. Then we set up the interrupt service routine. To do this you must enable the interrupt for each input in the register PCMSK0. Because we want the sketch to start in a well-defined state, we clear the interrupt flag in the register PCIFR (Pin Change Interrupt Flag Register) in case it had been set by a level change on one of the inputs that occurred while the sketch was busy configuring itself. Clearing this bit ensures that the first interrupt to be served will be due to a level change on a pin after this point. Finally, interrupt requests are enabled by setting bit PCIE0 of the PCIFR the register and the buffer is emptied.

The interrupt service routine `ISR(PCINT0_vect)` is called for each level change on one of the six enabled inputs. Since the MCU does not have a register to determine which pin caused the interrupt, we must provide a mechanism to do it ourselves. This is not terribly complicated; all you need is a variable to emulate the register. I called this variable `pcint0_states_prev` and you already met it in the function `setup`. The first thing that the interrupt service routine does is reading the inputs. To determine which one of them changed state, it performs a bitwise exclusive OR operation with the bits of the `pcint0_states_prev` "register" that contains the preceding states. Only the inputs that have changed level compared to their previous value will have a bit set to one in the variable `pcint`, the other bits are zero. Port B has eight pins of which only six are accessible. Because you may not have enabled all six inputs as I did here, and we therefore perform a bitwise AND operation to clear the bits of the unused inputs. Now we know which pins have changed level since the last time we read them. The current state of the inputs is stored in the variable `pcint0_states_prev` for the next time.

The action to take when a pin level changes depends on the application and can be different for each input. In this example all inputs write a different character in the data buffer. The routine ends by writing a separator character (a hyphen) in the buffer, which allows us to distinguish the different calls of the interrupt service routine in the serial monitor.

Let's finish this section with a little game: try, for example with two pieces of wire connected to ground, to generate two simultaneous level changes on two inputs. If you see two (or more) digits between two dashes, you succeeded. Well done![1]

10.8 Long Live the Rotary Encoder!

I love rotary encoders. Don't ask me why, but ever since I discovered, decades ago now, this kind of endless potentiometer; I am a fan. Rotary encoders can take the form of a "pot", but they are also found in the form of a disk with optical barrier. Furthermore, they are not always circular; they can also be straight (in which case it is referred to as an optical linear encoder). The most common type is the incremental encoder that provides a signal of which the level depends on the previous state or position. Absolute encoders provide an absolute angle. Today, rotary encoders are often used as endless dials in all kinds of equipment but, in robotics for example, they are also used for determining the position of an arm, the speed of a wheel or the direction of rotation of a motor.

The incremental rotary encoder is very simple: it produces pulses when it turns, like a disk with small holes in it that let light through when they pass in front of a light source (an LED for instance). A bike with a magnet attached to a spoke of the wheel to measure its speed is also a rotary encoder. The faster the encoder rotates, the higher the pulse rate. To measure rotational speed it is sufficient to measure the frequency of a pulse signal, but when the direction of rotation must also be determined, two signals are required. Two options present themselves:

- ✦ a direction signal together with a rotational speed signal;
- ✦ two signals phase-shifted by 90° (the most common solution) like a sine and cosine, but square.

---

1. If, like me, you did not succeed, know that you can cheat. Remember that the PCINT interrupts also work with outputs. If you change two inputs into outputs and if you change their levels simultaneously with `PORTB = xx` (it is up to you to determine the value of `xx`), two digits instead of one will be written into the buffer.

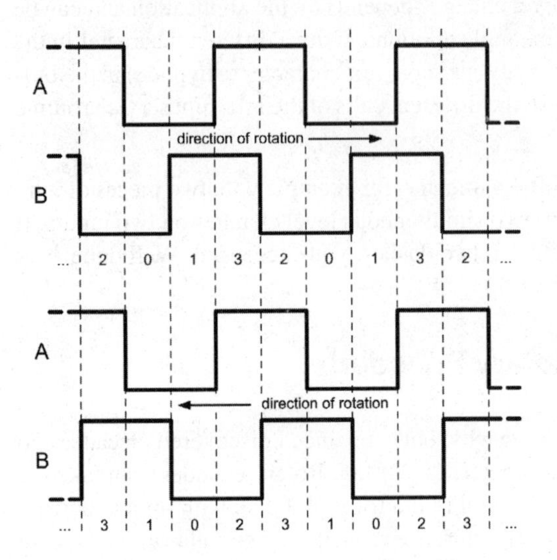

Figure 10-2 - The operating principle of a two-phase rotary encoder. When the encoder rotates in one direction, the signals A and B are phase-shifted by 90°; we might say that A is ahead of B. When the encoder turns in the other direction, B is ahead of A.

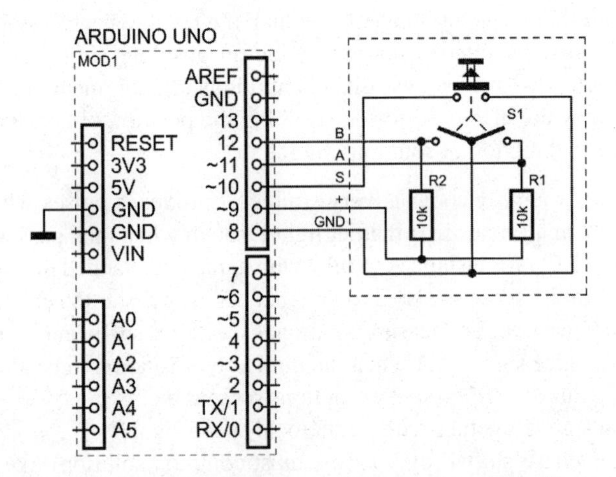

Figure 10-3 - My rotary encoder with integrated pushbutton together with two pull-up resistors are mounted on a small circuit board. Curiously, there was no pull-up resistor for the pushbutton.

In an application where the encoder is at rest most of the time, like the volume control of an audio amplifier or the controls on my oscilloscope (seven rotary encoders, I love it!), continuously scanning the inputs to which the encoder is connected, is a waste of time for the MCU. It is much better to use interrupts instead to reduce

Figure 10-4 - Close-up of the rotary encoder mounted on the Uno.

the attention that the MCU must give to the encoder to an absolute minimum. The level changes of the two signals A and B of the encoder are of equal importance, making PCINT-type inputs the perfect choice for reading such a device.

```
/*
 * rotary encoder
 */

int pin_gnd = 8;
int pin_vcc = 9;
int pin_sw = 10;
int pin_a = 11;
int pin_b = 12;
int state = 0;
int state_sub = 0;
volatile int counter = 0;
volatile boolean flag = false;

void encoder_read(void)
{
  uint8_t pin_states = PINB & PCMSK0;

  // pushbutton?
  if ((pin_states&0x04)==0)
  {
```

```
    counter = 0;
    flag = true;
  }

  pin_states = (pin_states>>3)&0x03;

  // pin_state is now 0, 1, 2 or 3.

  if (pin_states!=state)
  {
    // EXOR to determine de direction of rotation.
    int inc = ((pin_states>>1)^state)&0x01;
    if (inc==0) inc = -1;
    state = pin_states;

    // Optimisation in case of direction change.
    if ((inc<0 && state_sub>0) || (inc>0 && state_sub<0)) state_sub = 0;

    state_sub += inc;
    if (state_sub<=-4 || state_sub>=4)
    {
      state_sub -= (inc<<2);
      counter += inc;
      flag = true;
    }
  }
}

ISR(TIMER2_COMPA_vect)
{
  bitClear(TIMSK2,OCIE2A);       // Disable interrupt timer 2.
  encoder_read();
  bitSet(PCIFR,PCIF0);           // Clear pending interrupts.
  bitSet(PCICR,PCIE0);           // Enable PCINT0-7.
}

ISR(PCINT0_vect)
{
  bitClear(PCICR,PCIE0);         // Disable PCINT0-7.
  bitSet(TCCR2A,WGM21);          // CTC mode.
  bitSet(TCCR2B,CS20);
  bitSet(TCCR2B,CS22);           // Clock/128.
  OCR2A = 250;                   // 2 ms.
  TCNT2 = 0;                     // Clear the counter.
  bitSet(TIFR2,OCF2A);
  bitSet(TIMSK2,OCIE2A);
}

void setup(void)
{
  Serial.begin(115200);
```

```
  pinMode(pin_gnd,OUTPUT);
  digitalWrite(pin_gnd,LOW);
  pinMode(pin_vcc,OUTPUT);
  digitalWrite(pin_vcc,HIGH);
  pinMode(pin_sw,INPUT_PULLUP);
  pinMode(pin_a,INPUT);
  pinMode(pin_b,INPUT);
  state = 0;
  state_sub = 0;
  counter = 0;
  flag = false;
  PCMSK0 = 0x1c;
  bitSet(PCIFR,PCIF0);              // Clear pending interrupts.
  bitSet(PCICR,PCIE0);              // Enable PCINT0-7.
}

void loop(void)
{
  if (flag==true)
  {
    Serial.println(counter);
    flag = false;
  }
}
```

Unlike the previous example that used a buffer for communication between the interrupt service routine and the function `loop`, this time I used a variable `counter` with a variable `flag` to signal to the function `loop` that the value of `counter` was modified. If this is the case, its value is sent to the serial port. The variables `counter` and `flag` are both declared as **`volatile`**, even if this is not strictly necessary for `counter`. But since you never know how such a variable may be used in a real-life application, it is better to come prepared.

The function `setup` is relatively long, which is mainly due to the number of pins of the rotary encoder that I used. It is a model with an integrated pushbutton mounted on a small PCB with five contacts. This means that you must configure two outputs and three inputs. The common contact of the two phases A and B is called GND, so I connected it to an output set to 0 V. The pull-up resistors are connected to the PCB pin labeled '+' and my experiences have shown me that this pin had to be used to make the sketch properly detect level changes. This also explains why I have not activated the internal pull-up resistors for these inputs. The pushbutton does not have an external pull-up resistor and so I used the one built in the MCU.

After configuring the pins the function `setup` first initializes some global variables before enabling the interrupts on pins 10, 11 and 12. The procedure is the same as in the previous example. However, the interrupt service routine `ISR(PCINT0_vect)` has been completely remodeled. I did this for two reasons:

1. to create a debounce filter for the encoder contacts;
2. to show you a technique often used in microcontroller applications: interrupt chaining.

The contacts of my rotary encoder are mechanical, not optical, so they suffer from a phenomenon known as contact bounce. Mechanical contacts bounce several times before settling in their final position. One method to overcome this is to delay the reading of the contact until bouncing has stopped. To do this, instead of reading the contacts of the encoder, the interrupt service routine starts a timer. When the timer times out it generates an interrupt and it is in the routine that services the timer interrupt that the contact positions are read. I used timer/counter 2 with a delay of 2 ms for this, meaning that the encoder contacts are read in the second interrupt service routine of the sketch, `ISR(TIMER2_COMPA_vect)`, 2 ms after a level change on an input. The delay of 2 ms was chosen after a bit of experimentation. If the delay is too long, you risk missing level changes; if the delay is too short, the debounce filter is not very efficient.

The interrupt service routine `ISR(PCINT0_vect)` disables pin level change interrupt requests and then starts timer 2. This module is configured each time, because you never know what state it may be in. The interrupt service routine `ISR(TIMER2_COMPA_vect)` disables the interrupt requests of timer 2, it reads the contacts of the encoder, and then re-enables the PCINT interrupts. Doing things this way chains the two interrupts together. The function `loop` is not involved and has absolutely no idea of what is happening behind its back. Did I hear someone say multitasking?

The crux of the sketch is in the function `encoder_read`. It starts by reading the contacts, then it checks if the pushbutton is pressed. If this is the case, the variable `counter` is set to zero and `flag` is hoisted. Then the function examines the other two contacts to determine their states. Four states are possible, as shown in the tables below. The encoder must pass through all four states before the variable `counter` may be incremented or decremented. My rotary encoder provides tactile feedback when you spin it; each click completes a sequence of four states. If the state has changed since the last time (stored in the variable `state`), the function will attempt to determine the direction of rotation. The state sequence allows to determine the direction of rotation, but it is more elegant – and less comprehensible – to "exclusive OR" the two signals A and B, one of which delayed by one state. The columns labeled A* ^ B in the tables below show the result.

A	B	State	A*	A* ^ B
0	0	0	0	0
1	0	2	0	0
1	1	3	1	0
0	1	1	1	0
0	0	0	0	0

A	B	State	A*	A* ^ B
0	0	0	1	1
0	1	1	0	1
1	1	3	0	1
1	0	2	1	1
0	0	0	1	1

Table 10-3 - The sequences in the State columns show the direction of rotation of a rotary encoder with two phases A and B: 02310 versus 01320. However, a trick allows us to obtain a single bit value that indicates the direction of rotation. By delaying one of the two phases by one state (see the columns A*) before doing an exclusive OR ('^') with the other, non-delayed phase (B), we obtain a signal of which the level depends on the direction of rotation.

Once we know in which direction the encoder turns, we can start thinking about updating the variable `counter`, but, as I said earlier, only after the four states have been visited. The visited states are stored in `state_sub` and kept for the next round. Whenever the variable `state_sub` reaches the value of plus or minus 4, the variable `counter` is updated and the flag is raised to tell the function `loop` that the variable `counter` has received a new value.

10.9 Reset In Every Possible Way

I said earlier that interrupt INT0 has the highest priority of al interrupts. Well, that is not true. Even though the vector tables from Section 10.2.1 (page 290) start counting at one, there is also a vector zero. This missing vector is reserved for the reset signal that functions as an interrupt. For many MCUs the reset vector either points to the beginning of the memory (address 0) or to the end (address 0xffff…, the exact number of 'f's depends on the size of the memory), but not necessarily. The reset signal differs from interrupts because it causes more than the execution of an interrupt service routine, most of the MCUs registers are also reinitialized. Some MCUs have registers that are not affected by the reset signal and the only way to reinitialize them is by switching the power to the chip off and then on again. Note that resetting the chip does not mean that all the registers are cleared and that all the bits are set to zero (BTW, in some MCUs the default bit value is one). Reset means "to set again", which is not the same thing as "to clear" even though the word is often used in that sense. You have to read the MCU's user manual to know the exact value of a register after a chip reset.

10.9.1 POR, BOR and BOD

A microcontroller can have multiple reset signals. Powering the chip often (not always!) triggers a power-on reset or POR signal. Power sags or brownouts may form another source of reset signals, the brown-out reset or BOR. A special circuit, the Brown-Out Detector or BOD triggers this reset signal. In the AVR this feature is controlled by one of the fuse or configuration bits. The external reset is the best known source, because it is often connected to a push button. A fourth source of reset signals is the watchdog, a special timer which – when activated – triggers a restart of the MCU if the program does not tickle the watchdog periodically. It is like a duel, the MCU must reset the watchdog before the watchdog resets the MCU.

The microcontroller on the Mega board offers a fifth reset source: the JTAG port. Since the JTAG port allows an external device to take control of the MCU to for example reprogram the flash memory or to debug the program, the port must also have access to the reset signal. The MCU of the Uno board does not have a JTAG port.

When an MCU features several reset sources a special register is usually available so that the program can find out why it was restarted. On the AVR this register is called MCUSR and it contains a flag for each reset source. This register is used by the Arduino bootloader to determine if it must start the sketch immediately or if it should instead monitor the serial port because the arrival of a new sketch to be programmed into the flash memory is imminent. Watch the on-board LED (not the power indicator, of course), and switch on the board. The LED does not blink. Now press the reset button on the board while observing the LED. It will flash briefly three times if your board is a Uno programmed with the Optiboot bootloader, which is probably the case. My Mega board shows the same behavior in both cases, which makes it a bad example.

The reset function initializes the microcontroller to a higher level: it prepares the memory and the stack (see above), and it loads the global variables with their initial values, etc. When done, it will call the function `main` of the sketch to start the program. Now the peripherals and the application itself are initialized after which it can start.

10.10 Let's Switch Roles

So far this chapter explained how to interrupt the microcontroller by for example an event or a user action. What if we reversed the roles? What if we let the MCU interrupt its user? Often a system must attract the attention of an operator or supervisor to solve a problem, to respond to an alarm, to activate something or to provide information. The remainder of this chapter deals with this important topic.

10.10.1 The Annoiser

Do you know the prank of the gadget that, hidden in a dark room like a bedroom, emits a high-pitched beep at random times? This gag exploits the difficulty of the human ear to pinpoint a sound source accurately. When the sound occurs only in the dark and, in addition, at unpredictable times, the exercise becomes even more difficult. Just imagine the effect if the subject of this shenanigan is half asleep. Such an annoying beep is as worse as a swarm of mosquitos and you can expect the victim to pull a long face in the morning.

So now you want to know how to make such a device, right? Well, I am not going to tell you. Not only because it is not very original as a prank, but also because you should now be able to develop a suitable device yourself. No, no, do not insist, it is *nyet*.

Okay, but quickly then.

You will need a piezoelectric buzzer and a darkness detector. A buzzer is the perfect device for producing annoying sounds. Note that there are two types, one with a built-in oscillator and one without. The first type will produce a sound as soon as it is powered (check the polarity), but at a fixed frequency. It is possible to modulate the frequency, but if you do that the sound becomes less pure. The second type of buzzer must be driven with a periodic signal, a square wave for example, and has the advantage of being able to produce a range of frequencies, although they tend to have a preference for frequencies around two to four kilohertz. This type has no polarity despite the '+' that is often printed on it. To avoid problems and malfunctions, I advise you to Always Respect the Polarity[1].

To detect darkness we are going to use a light sensor. For this I propose a resistive voltage divider where one of the resistors is replaced by a photoresistor or Light-Dependent Resistor (LDR). The value of the photoresistor depends on the incident

1. This is one of the credos of the Arduino team. It is visualized by the '+' and '–' symbols in the Arduino logo.

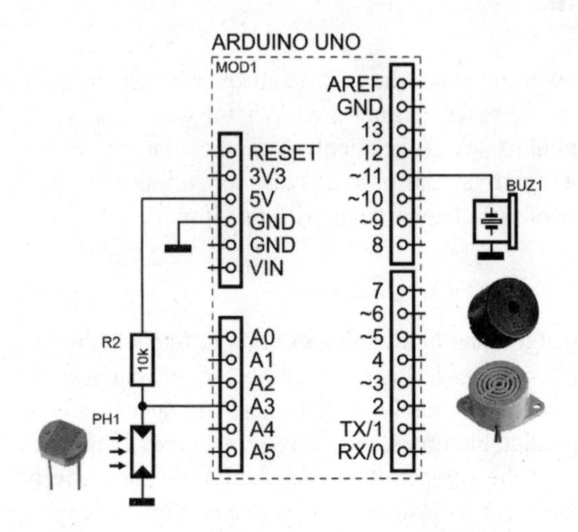

Figure 10-5 - Three components and an Arduino board can even deprive Rip van Winkle of sleep.

light. The more intense the light is, the lower its resistance will be. In total darkness, the photoresistor reaches values measured in megohms (MΩ), in full light its resistance can drop to less than a kilohm (kΩ).

The hardest part of the Annoiser is to produce sonic events at random times. To achieve this we will use the light sensor and a slightly strange calculation to make the pseudorandom number generator more random. Here is the sketch:

```
/*
 * buzzer 1
 */

#define BEEPS_MAX  7
#define DARK   900

int buzzer = 11;
int ldr = A3;
unsigned long int seconds;
int seconds_max;
int beeps;
int frequency[BEEPS_MAX];
int pulse[BEEPS_MAX];
int pause[BEEPS_MAX];

void arm(void)
{
  seconds = 0;
  seconds_max = random(600,3600);
```

```
  beeps = random(1,BEEPS_MAX);
  for (int i=0; i<beeps; i++)
  {
    pulse[i] = random(100,500);
    pause[i] = random(500,1500);
    frequency[i] = random(4000,5000);
  }
}

void setup(void)
{
  randomSeed(analogRead(ldr));
  delay(1000);
  randomSeed(random(1000000)*analogRead(ldr));
  arm();
}

void loop(void)
{
  delay(1000);
  if (analogRead(ldr)>DARK)
  {
    seconds += 1;
    if (seconds>=seconds_max)
    {
      for (int i=0; i<beeps; i++)
      {
        tone(buzzer,frequency[i],pulse[i]);
        delay(pulse[i]+pause[i]);
      }
      arm();
    }
  }
}
```

The function `loop` is slowed down to once per second with the function `delay`. Every second the sketch checks if it is dark outside and if it is, it verifies if the countdown of the random period is over. When this is the case, the buzzer is turned on. To spice up the sketch a bit, I not only made the interval between the sonic events random, but also the number of beeps, the beep frequencies, their duration and the pauses between the beeps. Note how, after calling the function `tone`, the function `loop` waits `pulse` + `pause` milliseconds instead of just `pause` milliseconds. The explanation for this can be found in Section 9.15.2, page 274.

At the end of the beep sequence, all the sonic event parameters are recalculated by the function `arm`.

The darkness detector simply measures the voltage at its input. If it surpasses a certain threshold (`DARK` in the sketch), the detector concludes that it is dark.

The only interesting thing in the function setup is perhaps the initialization of the pseudorandom number generator. For this I also used the photoresistor. It is read twice at a one second interval so that, with a little luck, two slightly different values are obtained. The first one initializes the generator (randomSeed) and a random value is produced with a call of the function random. The random value is multiplied by the second reading of the photoresistor and the result is used to reinitialize the pseudorandom number generator. This method seems to work pretty well due to ever present noise, even under constant ambient light conditions.

The function arm calculates random values for all parameters of the alarm.

That's it, your Annoiser is ready. Now you have to find a victim. I advise against testing it on your spouse for two reasons. Firstly because earplugs do not block the buzzer beeps too well and therefore you will suffer as much as your spouse, and secondly because your spouse may disallow you to spend any of your free time on programming microcontrollers.

10.11 La Cucaracha

Although the previous sketch is fun, I think it is possible to do better. For example, instead of generating tones with random frequencies and durations, the sketch could play a little tune. It is easy to do so, just fill the frequency, pulse and pause tables with the right values. Here's how to modify the previous sketch to make it play (massacre?) the first few notes of a sonata composed by Domenico Scarlatti:

```
#define BEEPS_MAX  7
int beeps = BEEPS_MAX;
int frequency[BEEPS_MAX] = { 2960, 2793, 2637, 3951, 2489, 3729, 3951 };
int pulse[BEEPS_MAX] = { 100, 100, 100, 100, 100, 100, 200 };
int pause[BEEPS_MAX] = { 50, 50, 150, 150, 150, 150, 150 };

void arm(void)
{
  seconds = 0;
  seconds_max = random(600,3600);
}
```

Manually filling the tables with frequencies, durations and pauses is tedious, which is why I looked for a preprogrammed component that could spare me this boring work. Now since Arduino is of Italian origin, what would be more natural than to add an Italian-style[1] musical car horn to it? The thing I am referring to is an air horn with five – or two or three – horns and that plays La Cucaracha or another popular

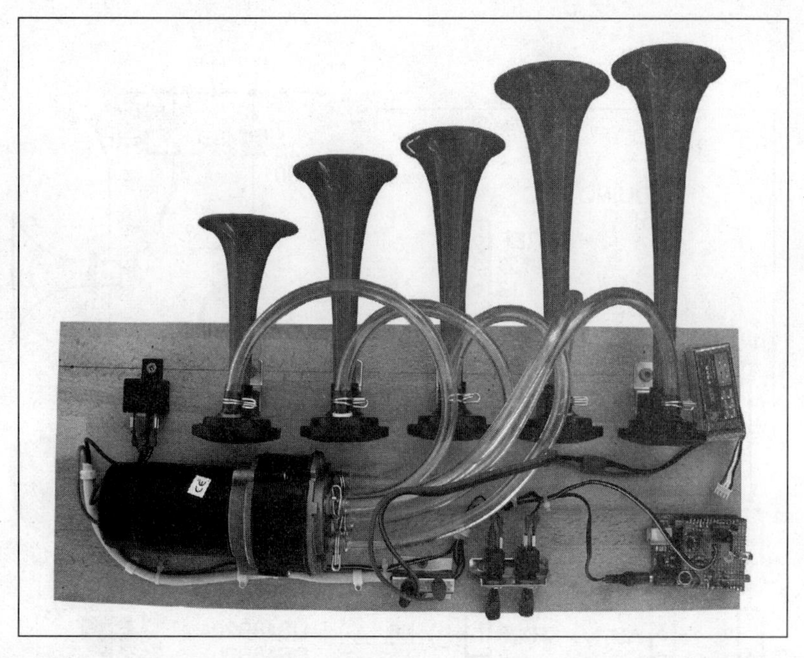

Figure 10-6 - A small air compressor and five horns can wake up an entire neighborhood. The battery capable of powering this contraption is the small block on the right, next to the longest horn.

tune at the quite adequate sound level of 115 dB, not far from the pain threshold. You know what I mean, you have heard them before. Equipped with such a device, we can build an effective alarm system or clock.

The idea is to install the alarm system outside, hidden in a tree or in the bushes, where it waits until the night falls and the temperature drops enough to activate the "ringtone". Unlike the previous sketch, this time the moment to trigger the alarm is not random, but depends on the ambient temperature. Thus, early in the morning, when it is coolest, the alarm goes off and wakes up the whole neighborhood, especially in summer when people tend to sleep with their windows open. The first few times your neighbors will put a pillow on their heads, curse the inconsiderate honking driver and try to go back to sleep, but after a few of these interrupted nights they will become suspicious and start asking questions. Once you're exposed, you

1. Made popular by the Italian company FIAMM, one of the world's largest horn manufacturers.

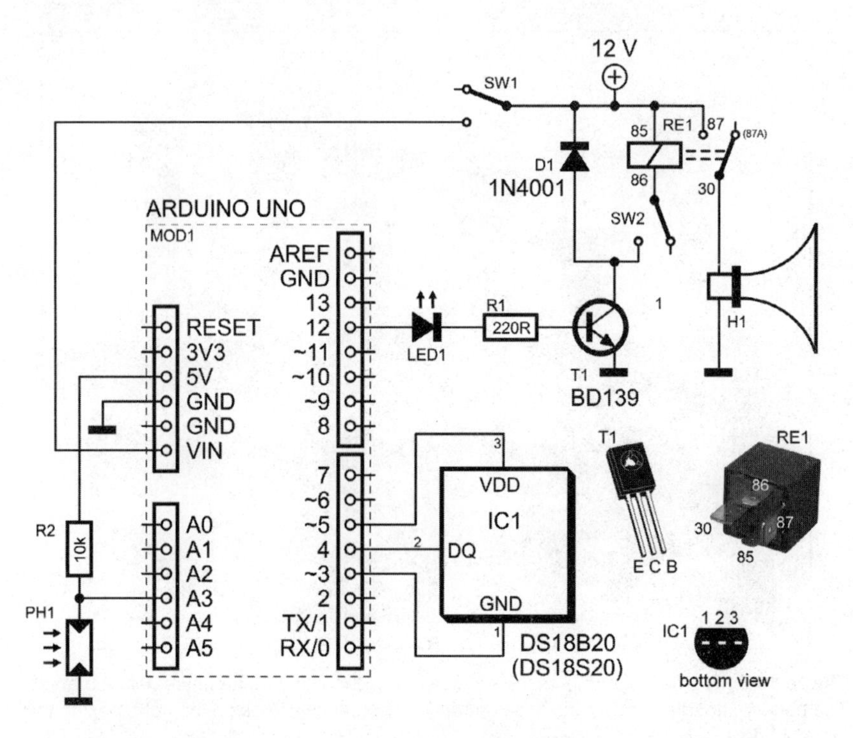

Figure 10-7 - La Cucaracha alarm. A 1-Wire temperature sensor and a light detector determine when the five tone horn will go off. The switches are useful to test different parts of the circuit without hurting your ears.

have a fair chance of being expelled from the neighborhood or village, but that does not really matter, because it will allow you to spend all your free time again on programming microcontrollers. So, how do we go about?

The system consists of a small air compressor powered at 12 V with five outputs and as many horns connected with flexible tubes to the compressor. The compressor contains a disk with small holes in it that enable the outputs of the compressor in a predetermined order to play a short melody. It is a kind of barrel organ, a mechanical musical wind instrument. You might even compare it to a rotary encoder.

Powering this musical instrument is not so simple, because it consumes a lot of energy. I measured about 18 A at 12 V, i.e. more than 200 watts. If you don't mind investing in a car battery, providing the required power does not pose too many problems. However, such a battery is expensive and quite heavy, which has a negative impact on the portability of the system. Fortunately, I found a cheap solution in the shape of a Lithium polymer battery also known as LiPo battery. These little wonders, used extensively in remote controlled planes, boats and cars, are commercialized with different output voltages – but always a multiple of 3.7 V – and with all kinds of capacities. I bought one for about ten dollars with an output voltage of 11.1 V (the closest I could get to 12 V) and a capacity of 1200 mAh. This battery, measuring 7 × 2 × 3 cm and weighing only 90 grams, is a "25C", which means that it is capable of delivering a peak current equal to 25 times its rated capacity. For my battery this corresponds to 30 A at 11.1 V, which is more than 300 watts! Its only drawback is that you need a special charger/balancer to recharge it.

The alarm came with a relay (for cars, rated 30 A at 12 V), a good thing because the microcontroller alone is unable to switch the required power. A transistor is needed for the MCU to drive this relay that consumes about 250 mA at 12 V. I therefore selected a transistor capable of switching this much current without getting hurt. An LED connected between the MCU and the base of the transistor provides visual feedback and shows if the relay is active or not.

10.11.1 The 1-Wire Protocol

For the temperature sensor I picked the ever popular DS18S20 or DS18B20 from Maxim. It's a sensor that communicates via a 1-Wire bus of which Maxim has become the owner when it took over its inventor Dallas Semiconductor. Because this bus appears on its way to obsolescence – the manufacturer stopped production of most of its 1-Wire components – we are not going to develop a generic 1-Wire library for Arduino. If you need one, search the internet where you will easily find what you are looking for.

The peculiarity of the 1-Wire bus is that is uses only one wire (not counting signal common), hence its name. Some of the compatible devices not only communicate over this wire, they can also be powered by it. It's a bit as if they were talking with their mouths full. Furthermore, several devices can be connected to the same wire to create a network or a bus with one master and several slaves. Each device has a unique 64-bit address that allows you to select it. Despite these quite impressive possibilities, it seems that the 1-Wire bus never really caught on. So, since the bus may soon become extinct, we will only write enough functions to read the temperature sensor and leave it at that.

When idle, the level on the bus is high due to a pull-up resistor. To send a logic zero, the master or the slave pulls the line low; to transmit a logic one, the line is disconnected and pulled high by the pull-up resistor. The master always initiates

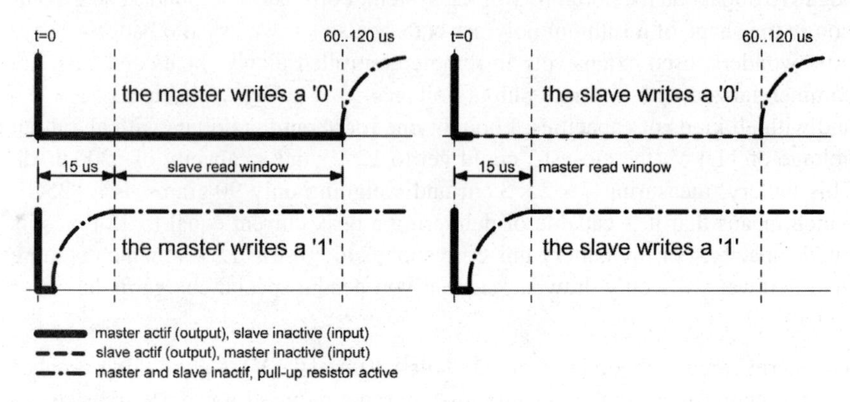

Figure 10-8 - In theory the 1-Wire protocol recognizes four different states, in reality there are only three, but an observer will see no more than two. No matter how many states there are, only the master is allowed to start transactions.

communications. It must first wake up the slave device – the sensor – by sending a reset signal in the shape of a 500-microsecond pulse. The sensor responds by pulling the line low for a not so very well defined duration, but never longer than 240 microseconds. Despite this, the master is supposed to wait at least 480 microseconds before continuing. Then the master sends the command to start a temperature measurement and waits until the sensor has finished. Before it can send the read command and read the data the master must again restart the sensor. The master signals the end of the communication by restarting the sensor once more and then hangs up on it.

Transmitting one data bit takes 60 µs – transferring a 64-bit value therefore takes 3840 µs –, the three restarts together eat up 3 ms and the necessary commands another 1920 µs, adding up to almost 9 ms for any single transaction. When there are several devices on the bus, the master must also send the 64-bit addresses of the slaves, which does not help to speed up the communication. This is what it looks like as a sketch:

```
/*
 * cucaracha
 */

#define DARK  1000
#define TEMPERATURE_MIN  16
#define REPEAT_MAX  2
```

```
int ds1820_gnd = 5;
int ds1820_vdd = 3;
int relay = 12;
int ldr = A3;
int repeat;

#define DS1820_DQ_DDR  DDRD
#define DS1820_DQ_PORT  PORTD
#define DS1820_DQ_PIN  PIND
#define DS1820_DQ_BIT  4
#define DS1820_SCRATCHPAD_SIZE  9
uint8_t scratchpad[DS1820_SCRATCHPAD_SIZE];

#define DS1820_DQ_LO DS1820_DQ_DDR |= _BV(DS1820_DQ_BIT); \
                     DS1820_DQ_PORT &= ~_BV(DS1820_DQ_BIT)

#define DS1820_DQ_HI DS1820_DQ_PORT |= _BV(DS1820_DQ_BIT); \
                     DS1820_DQ_DDR &= ~_BV(DS1820_DQ_BIT)

#define DS1820_DQ_IN (DS1820_DQ_PIN & _BV(DS1820_DQ_BIT))

boolean ds1820_reset(void)
{
  boolean presence = false;
  DS1820_DQ_LO;
  delayMicroseconds(500);
  DS1820_DQ_HI;
  int timeout = 480;
  int dt = 30;
  while (timeout>0)
  {
    delayMicroseconds(dt);
    timeout -= dt;
    if (DS1820_DQ_IN==0)
    {
      presence = true;
      break;
    }
  }
  // Finish timeout.
  delayMicroseconds(timeout);
  return presence;
}

uint8_t ds1820_time_slot(uint8_t value)
{
  uint8_t result = 0;
  DS1820_DQ_LO;
  delayMicroseconds(2);
  if (value!=0) DS1820_DQ_HI;
  delayMicroseconds(15);
```

```
  if (DS1820_DQ_IN!=0) result = 1;
  delayMicroseconds(45);
  DS1820_DQ_HI;
  return result;
}

void ds1820_write_byte(uint8_t value)
{
  for (uint8_t mask=0x01; mask!=0; mask<<=1)
  {
    ds1820_time_slot(value&mask);
  }
}

uint8_t ds1820_read_byte(void)
{
  uint8_t result = 0;
  for (uint8_t mask=0x01; mask!=0; mask<<=1)
  {
    if (ds1820_time_slot(1)!=0) result |= mask;
  }
  return result;
}

boolean ds1820_read_temperature(void)
{
  boolean result = false;
  if (ds1820_reset()==true)
  {
    ds1820_write_byte(0xcc);
    ds1820_write_byte(0x44);
    while (ds1820_time_slot(1)==0);
    if (ds1820_reset()==true)
    {
      ds1820_write_byte(0xcc);
      ds1820_write_byte(0xbe);
      for (int i=0; i<DS1820_SCRATCHPAD_SIZE; i++)
      {
        scratchpad[i] = ds1820_read_byte();
      }
      result = true;
      ds1820_reset();
    }
  }
  return result;
}

void setup(void)
{
  pinMode(ds1820_gnd,OUTPUT);
  digitalWrite(ds1820_gnd,LOW);
```

```
  pinMode(ds1820_vdd,OUTPUT);
  digitalWrite(ds1820_vdd,HIGH);
  pinMode(relay,OUTPUT);
  digitalWrite(relay,LOW);
  repeat = 0;
}

void loop(void)
{
  delay(1000);

  ds1820_read_temperature();
  float t = (scratchpad[1]*256.0 + scratchpad[0])/16.0;

  if (analogRead(ldr)>DARK)
  {
    if (t<TEMPERATURE_MIN && repeat<REPEAT_MAX)
    {
      digitalWrite(relay,HIGH);
      delay(3000);
      digitalWrite(relay,LOW);
      for (int i=0; i<5; i++)
      {
        delay(60000);
      }
      repeat += 1;
    }
  }
  else
  {
    repeat = 0;
  }
}
```

The function `loop` is clocked at 1 Hz and the temperature is measured during each pass. Note that the calculation concerns the 12-bit DS18***B***20 temperature sensor, for the 9-bit DS18***S***20 you must divide the sum by two instead of sixteen.

If it's dark outside with the temperature below a certain value, and if the alarm has not sounded often enough, the compressor is turned on. My horn needs about three seconds to play its sweet melody; the relay is therefore activated for three seconds. Then the sketch performs a five-minute interval. This recess is timed using a loop that counts minutes, so it is easy to adjust the interval in minutes. Finally, the repeat counter is incremented so that the alarm will repeat a few times, but not too many.

The relay control output and the temperature sensor pin are initialized in the function `setup`. Nothing complicated, you know the ropes by now.

The function ds1820_read_temperature embodies the temperature reading algorithm that I described earlier. There are a few extra instructions like two calls of ds1820_write_byte(0xcc) that allows us to skip reading the sensor's read-only memory. Normally you have to read all of the sensor's memory, ROM and RAM, but by issuing this special command you can skip the ROM. We read all nine bytes of the RAM, the so-called scratchpad, because it is not possible to read a single byte. The first two bytes contain the temperature, we can ignore the others.

The bits of the bytes are sent and read using two very similar functions ds1820_write_byte and ds1820_read_byte. Both use the same function ds1820_time_slot, since for the master signalling a bit-read condition or writing a logical one is the same thing: pulling the line low for a short time before disconnecting the output. In the case of a read operation, the master must wait 15 µs before reading the input. In the case of a write operation, nothing forbids the master to read the pin after 15 µs, regardless of whether it is useful or not. In the case of a write operation, the value read can simply be ignored. In short, looking at things this way it becomes possible to combine reading and writing of the bus in a single function.

The function ds1820_reset handles the bus reset signal.

The only difficulty of the 1-Wire protocol is respecting the pulse and space lengths. Because of that, I did not use the functions pinMode, digitalRead and digitalWrite because they are quite slow. Instead I created the macros DS1820_DQ_HI, DS1820_DQ_LO, DS1820_DQ_IN that only do the strict minimum: configure a port as an input or output before reading or writing it. Note that writing a logic one is achieved by configuring the pin as an input with a pull-up resistor, as specified by the 1-Wire protocol. This is the pull-up resistor that sets the level of the bus.

10.12 Fire!

Some time ago lightning struck my neighborhood. I did not have much damage, but the impact did destroy my internet modem, my wireless router and an external hard drive, two devices that were connected to the modem by Ethernet cables. The lightning caused a strong voltage transient that entered the house through the telephone line. The phone itself was not affected, but all the other devices connected directly or indirectly to the phone line suffered severe overvoltage damage resulting in burnt components. They were installed together on a shelf in my bedroom, surrounded by books. Fortunately – relatively speaking – lightning struck just before I went to bed so I immediately noticed the smell of burning and was able to avoid the worst. I installed my new modem in the living room, away from flammable objects.

In Europe, it is estimated that a home fire starts about every two minutes. According to emergency service statistics 30% of home fires are of electrical origin, which amounts to about 240 home fires per day caused by an electrical problem. A short circuit, a device or a cable that overheats or, like above, a lightning strike, you are never safe from fire. Installing smoke detectors is a good idea, but they are a bit slow. When the motor of a home ventilation system overheats, it takes a little time before the smoke particles reach the smoke detector. Why not try to make a faster smoke detector that triggers an alarm as soon as a (motor) temperature rises dangerously? Every second that we can win may result in a home saved. So, how can we go about?

10.12.1 The SMBus

It is yet again a temperature sensor that we will set to work, but this time a non-contact type. Why? Because not only are these sensors very fast, they can also monitor an area rather than a specific point. Indeed, if we put a non-contact sensor at a certain distance of a device susceptible to overheating, a motor for instance, the sensor can have a view of the entire device and monitor the temperature of the entire object, instead of the small aera the sensor happened to be stuck on.

For a non-contact temperature sensor I chose the MLX90614 from Melexis. This is an infrared radiation detector that is part of a whole family. The model used here is a sensor that operates at 5 V, measures temperatures up to 385 °C (725 °F) and communicates over a System Management Bus (SMBus, developed by Intel). To use the sensor, simply connect it to the MCU, that's all there is to it. The sketch is also quite simple as the SMBus protocol is basically[1] a simplified I²C or TWI protocol and we can use the Arduino Wire library we already used in Chapter 8 to communicate with the atmospheric pressure sensor HP03S. The biggest difference between the SMBus and the I²C bus is a Packet Error Checking or PEC byte which contains a checksum covering the entire frame. In our application this sum is calculated by the slave (the sensor) over all the bytes including the ones sent by the master. Calculating the checksum is similar to the SHT11 sensor from Chapter 8 but does not use the same polynomial ($x^8 + x^2 + x + 1$ for the MLX90614, the SHT11 uses $x^8 + x^5 + x^4 + 1$, please refer to the literature to figure out what this

1. To be true, the SMBus is a derivative of I²C, not just a simplification. For the purposes of our example however we can consider it to be a simplified I²C bus.

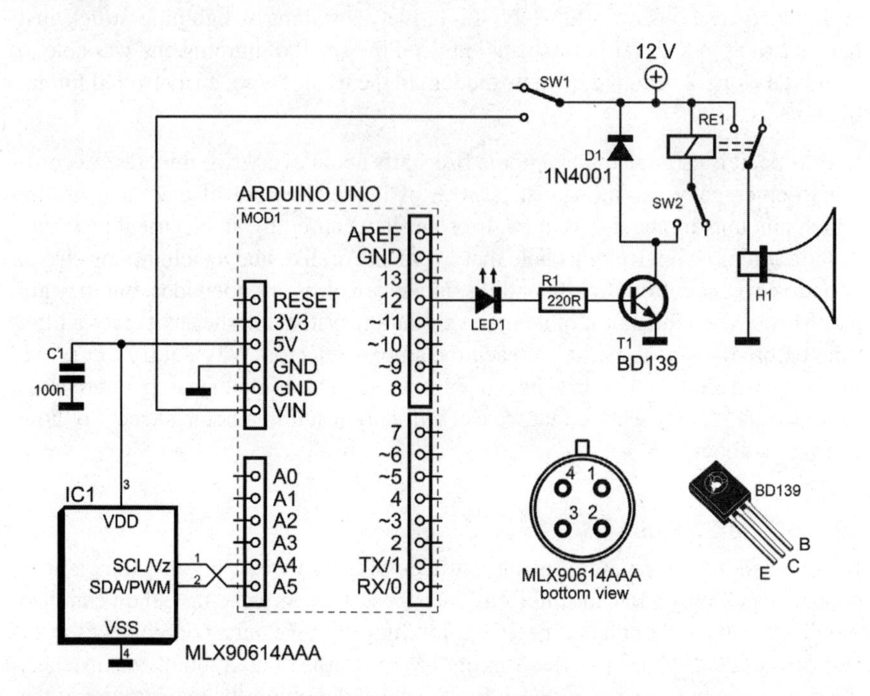

Figure 10-9 - Our fire detector is based on an infrared radiation sensor, i.e. a non-contact temperature sensor. Beware, IC1 is also available as a 3.3 V device, known as MLX90614***B***AA

means). This causes the contents of the table used in the calculation to differ. Chapter 8 does not show this table, but you can find it in the sketch that you can download. I decided to include the table for the SMBus in this sketch, because we are talking about an alarm that is not allowed to go off unnecessarily. We want to be sure of the correctness of the received data and verifying the checksum can improve our confidence.

Now let's have a look at the sketch:

```
/*
 * fire detector
 */

#include <Wire.h>

#define TEMPERATURE_MAX  25
#define MLX90614_ADDRESS  0x5a
```

Figure 10-10 - The infrared radiation sensor, housed in a metal can with an opening in the top, is mounted in the center of my second experimentation shield. The relay driver is to the right of the large two-contact connector.

```
#define MLX90614_READ_TEMPERATURE  0x07

int alarm = 12;
int mlx90614_sda = A4;
int mlx90614_scl = A5;
uint8_t crc = 0;

const uint8_t MLX90614_crc_table[256] =
{
  0, 7, 14, 9, 28, 27, 18, 21, 56, 63, 54, 49, 36, 35, 42, 45,
  112, 119, 126, 121, 108, 107, 98, 101, 72, 79, 70, 65, 84, 83, 90, 93,
  224, 231, 238, 233, 252, 251, 242, 245, 216, 223, 214, 209, 196, 195,
  202, 205, 144, 151, 158, 153, 140, 139, 130, 133, 168, 175, 166, 161,
  180, 179, 186, 189, 199, 192, 201, 206, 219, 220, 213, 210, 255, 248,
  241, 246, 227, 228, 237, 234, 183, 176, 185, 190, 171, 172, 165, 162,
  143, 136, 129, 134, 147, 148, 157, 154, 39, 32, 41, 46, 59, 60, 53, 50,
  31, 24, 17, 22, 3, 4, 13, 10, 87, 80, 89, 94, 75, 76, 69, 66, 111, 104,
  97, 102, 115, 116, 125, 122, 137, 142, 135, 128, 149, 146, 155, 156,
  177, 182, 191, 184, 173, 170, 163, 164, 249, 254, 247, 240, 229, 226,
  235, 236, 193, 198, 207, 200, 221, 218, 211, 212, 105, 110, 103, 96,
  117, 114, 123, 124, 81, 86, 95, 88, 77, 74, 67, 68, 25, 30, 23, 16, 5,
  2, 11, 12, 33, 38, 47, 40, 61, 58, 51, 52, 78, 73, 64, 71, 82, 85, 92,
  91, 118, 113, 120, 127, 106, 109, 100, 99, 62, 57, 48, 55, 34, 37, 44,
  43, 6, 1, 8, 15, 26, 29, 20, 19, 174, 169, 160, 167, 178, 181, 188,
  187, 150, 145, 152, 159, 138, 141, 132, 131, 222, 217, 208, 215, 194,
  197, 204, 203, 230, 225, 232, 239, 250, 253, 244, 243
};
```

```
void MLX90614_crc_update(uint8_t value)
{
  crc = MLX90614_crc_table[value^crc];
}

uint16_t MLX90614_read(void)
{
  uint16_t value = 0;
  uint8_t pec = 0;
  uint8_t lsb;
  uint8_t msb;

  // Read temperature register.
  Wire.beginTransmission(MLX90614_ADDRESS);
  Wire.write(MLX90614_READ_TEMPERATURE);
  // Restart without sending a stop condition.
  Wire.endTransmission(false);
  Wire.requestFrom(MLX90614_ADDRESS,3,true);
  if (Wire.available()!=0) lsb = Wire.read();
  if (Wire.available()!=0) msb = Wire.read();
  if (Wire.available()!=0) pec = Wire.read();
  Wire.endTransmission(true);

  crc = 0;
  MLX90614_crc_update(MLX90614_ADDRESS<<1);
  MLX90614_crc_update(MLX90614_READ_TEMPERATURE);
  MLX90614_crc_update((MLX90614_ADDRESS<<1)|0x01);
  MLX90614_crc_update(lsb);
  MLX90614_crc_update(msb);
  if (pec==crc) value = ((msb&0x7f)<<8) + lsb;
  return value;
}

void setup(void)
{
  Serial.begin(115200);
  pinMode(alarm,OUTPUT);
  digitalWrite(alarm,LOW);
  pinMode(mlx90614_sda,INPUT_PULLUP);
  pinMode(mlx90614_scl,INPUT_PULLUP);
  Wire.begin(MLX90614_ADDRESS);
}

void loop(void)
{
  delay(100);

  float t = MLX90614_read();
  t = t*0.02 - 273.15;
  Serial.println(t);
```

```
    if (t>TEMPERATURE_MAX)
    {
      Serial.println("alarm!");
      digitalWrite(alarm,HIGH);
      delay(3000);
      digitalWrite(alarm,LOW);
      delay(3000);
    }
}
```

I explained that the SMBus is a simplified version of the I²C bus; in the same vein the function `loop` is a simplified version of the function `loop` from the previous example, with a few differences. Here the function `loop` is clocked at 10 Hz instead of 1 Hz and it monitors the temperature, not the darkness. When the measured temperature exceeds a certain threshold, the alarm is triggered and La Cucaracha is played as loud as possible. As in the previous example, the goal here is to wake up everyone. Unlike the previous sketch, the alarm will now continue to sound until the temperature drops below the set threshold or until the power is turned off (a flat or burned battery for example).

The temperature calculation is different too. In fact, the MLX90614 sensor is extremely accurate if its ideal operating conditions are met – which is not necessarily the case here – and it is capable of measuring temperature with a precision of 0.02 °C (0.036 °F). Additionally, the sensor offers many possibilities to influence the processing carried out by its built-in signal conditioner, but the datasheet is so complicated that only the most persistent are able to benefit from this (certainly not me). Anyway, for this application the factory settings are suitable.

The function `setup` is so simple that I will not go into detail about it.

The temperature measurement is carried out in the function `MLX90614_read` that implements the read algorithm. After opening the hostilities, the command to read the temperature is sent. To continue from this point you need to restart the bus without creating a stop condition (see also Section 8.4.2, page 190), which is possible with the command `Wire.endTransmission(false)`. Then the master requests three bytes from the sensor, two temperature bytes and one checksum byte, and then hangs up. The checksum is calculated over all the bytes of the transaction, the address and read/write bits included, which explains the five calls to `MLX90614_crc_update` for only two useful data bytes. If the received checksum matches the calculated checksum, the function assembles the temperature from the two other data bytes before returning the value to the function `loop`.

Calculating the checksum is "classic" and simple thanks to the table. The latter takes up memory space, but this is not a problem in our application. The calculation is cumulative, which means that the global variable `crc` is updated each time the function `MLX90614_crc_update` is called to track the data flow of, in our case, five bytes. The table and the algorithm can both be found on the internet as they are part of the SMBus standard.

So, now that you are in the possession of a good fire alarm, put it into good use. I suggest you install it in your home and also in the homes of your friends. This way, if one day, somewhere one of them goes off, maybe a home and lives will be saved thanks to you and you will forever be the hero of your entourage. In the streets, people will shoot admiring glances at you; they will praise you and enrol their off-spring in the same schools as the ones you went to with the hope that maybe one day their kids will follow in your footsteps. Reverent gratitude will be the recom-pense for your friendly deeds. And you, Master of MCU, the people invoke you. Out of the helpless troubles of war, through your power, they look at the world in security[1]. And there you are, forever free to spend your time on programming microcontrollers because your passion finally enjoys the respect it deserves from your entourage.

1. Freely adapted from a poem for Hieron of Syracuse by the Greek poet Pindar (not to be confused with Pandore).

11. Circuits and Exercises

11.1 Introduction

After reading ten chapters on programming and being subject to electronics tendencies you now probably feel like picking up your soldering iron again. I did present some circuit diagrams for you to try out, but did you really do that? Some circuits were really simple, but did you take the time to mount the more complex ones on prototyping board? I suppose not, and I do not blame you, oh no, wiring a circuit on "perf board" is pretty tedious work indeed. For this reason I decided to design a printed circuit board (PCB) so that you can practice your new programming skills on real hardware in a comfortable way.

11.1.1 One Size Fits All

If you studied closely the designs presented in the previous chapters you may have noticed that they could be subdivided into groups where the circuits from every group could be built on one and the same prototyping board, nicely sharing the Arduino pins. This grouping of circuits was not accidental but due but to the way I developed the chapters, adding new examples to my test board until all the Arduino pins were used. When a board was full, I started a new one. This means that it is difficult to design a PCB that can hold all the circuits from this book, because different circuits have to share Arduino pins. In some cases this is possible, in other cases it is not.

The PCB that I came up with is as multifunctional as possible. It allows you to have a play with many of the designs from the previous chapters, although not all of them will work in exactly the same way. This is actually an advantage for me, the educator, as it offers me the possibility to present slightly different problems that you should now be able to solve. For you, the student, it is interesting too because you will discover some new techniques. However, I was obliged to make some compromises. The result is that the circuits from Chapter 6 are not supported by the PCB. The H-bridge motor driver from Chapter 7 cannot be mounted on the PCB, but the experiment is still possible. Finally, I discovered that the pressure sensor HP03S with I²C interface from the manufacturer Hope RF as used in Chapter 8 is either no longer on the market or very difficult to obtain and so I have replaced it by a different pressure sensor, analogue this time.

11.1.2 Here We Go!

The schematic of the multipurpose PCB is drawn in Figure 11.1. This drawing shows all the components that can be mounted on the PCB, or, to be more exact, that have a footprint drawn on the PCB. As you will see in what follows, we will not always respect the footprint but use the provided pads for something else. You are not supposed to mount all the components together on the PCB – this is not even possible as some components share mounting pads – but if you do, rest assured, nothing will go up in smoke.

To use the multipurpose PCB with an Arduino board you must at least mount the pin headers K2, K3, K4 and K5. The pin headers are meant to be mounted on the bottom (or copper) side, all the other components should go on the top (or component) side.

I will now describe the different circuits that can be mounted on the multipurpose PCB in the order as they appeared in the previous chapters. Due to PCB design compromises the complexity does not increase gradually, but is distributed rather randomly.

11.2 LED Dimmer

Inspired by Figure 7-3 (page 123) this little circuit requires only three components: P1, R1 and LED1. The corresponding sketch is listed in Section 7.2 (page 122). But, beware! This sketch will not work without a small modification. Have a good look at Figure 11-2, and then look carefully at Figure 7-3. Notice the difference? Indeed, you must change the pin number of the slider of P1. I will not tell you how to do this; you should be able to figure it out all by yourself. Refer to Section 7.2 if you are not sure of your solution.

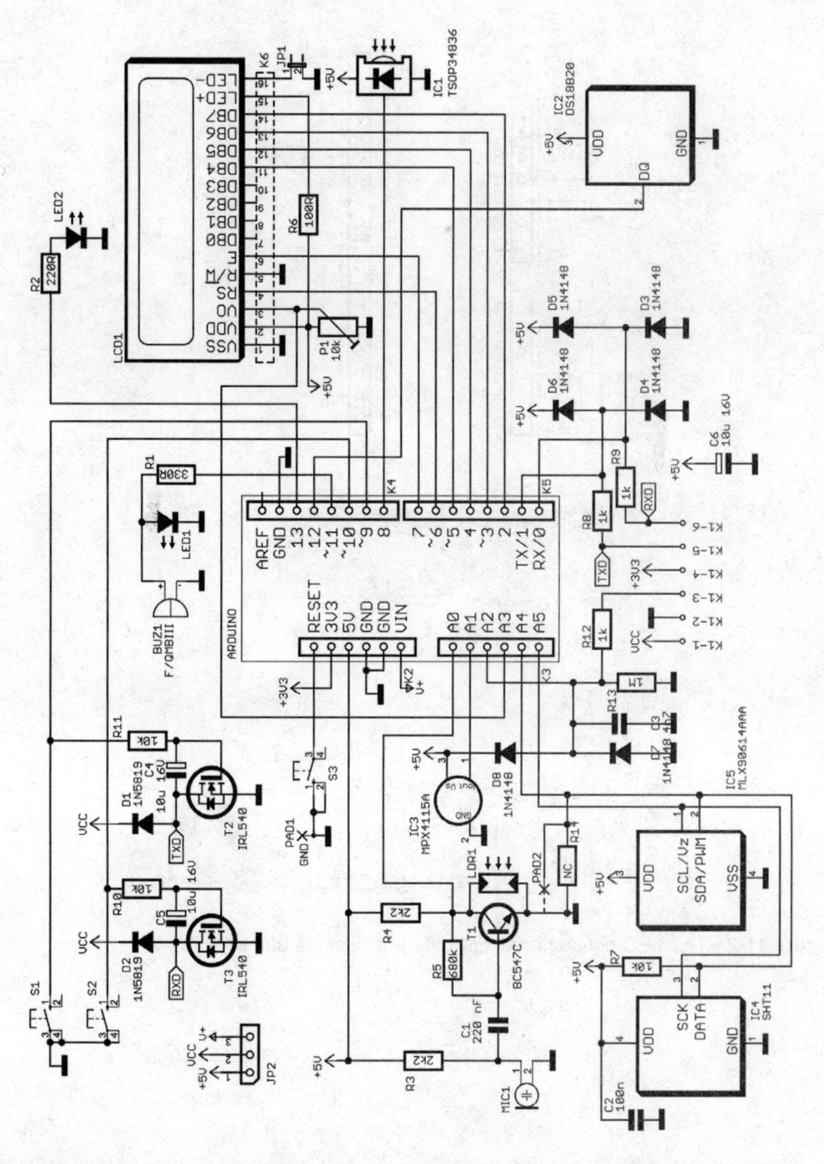

Figure 11-1 - The schematic of the multipurpose PCB is rather dense due to the multitude of possible circuits. Do not mount all the components on the PCB at the same time even though this should not create real problems. The values of most – if not all – resistors depend on the application.

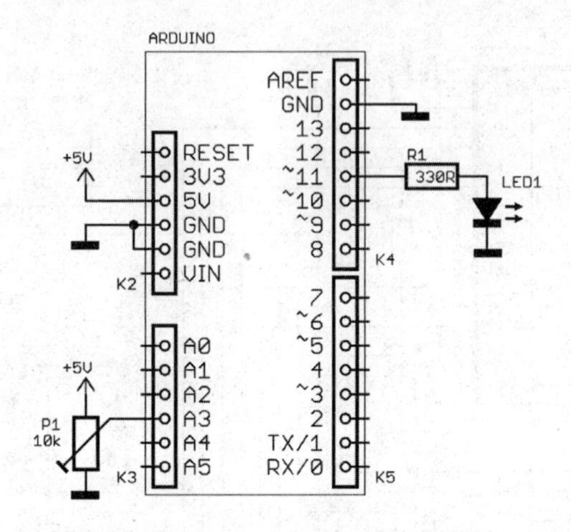

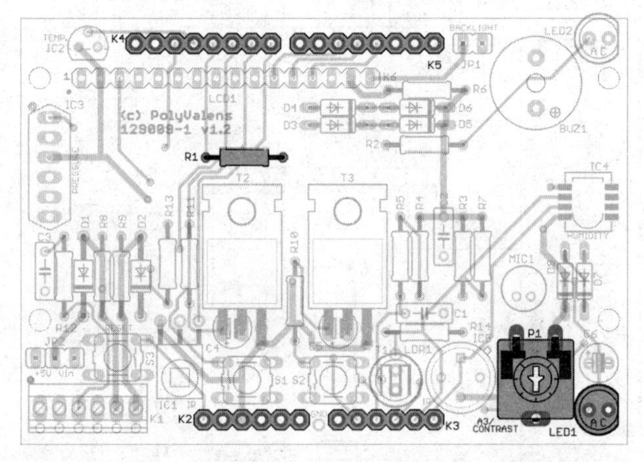

Figure 11-2 - This LED dimmer is slightly different from the one in Figure 7-3.

11.3 Motor Driver

The motor driver from Figure 7-5 (page 126) is an H-bridge using four big transistors plus two little ones and offering full control over the motor's rotation speed and direction. This circuit cannot be mounted on the multipurpose PCB. However, I did put two power MOSFETs on the PCB and they are connected to the terminal block K1. By adding an external double-pole double-throw (DPDT) relay we can make a bidirectional motor driver out of these transistors.

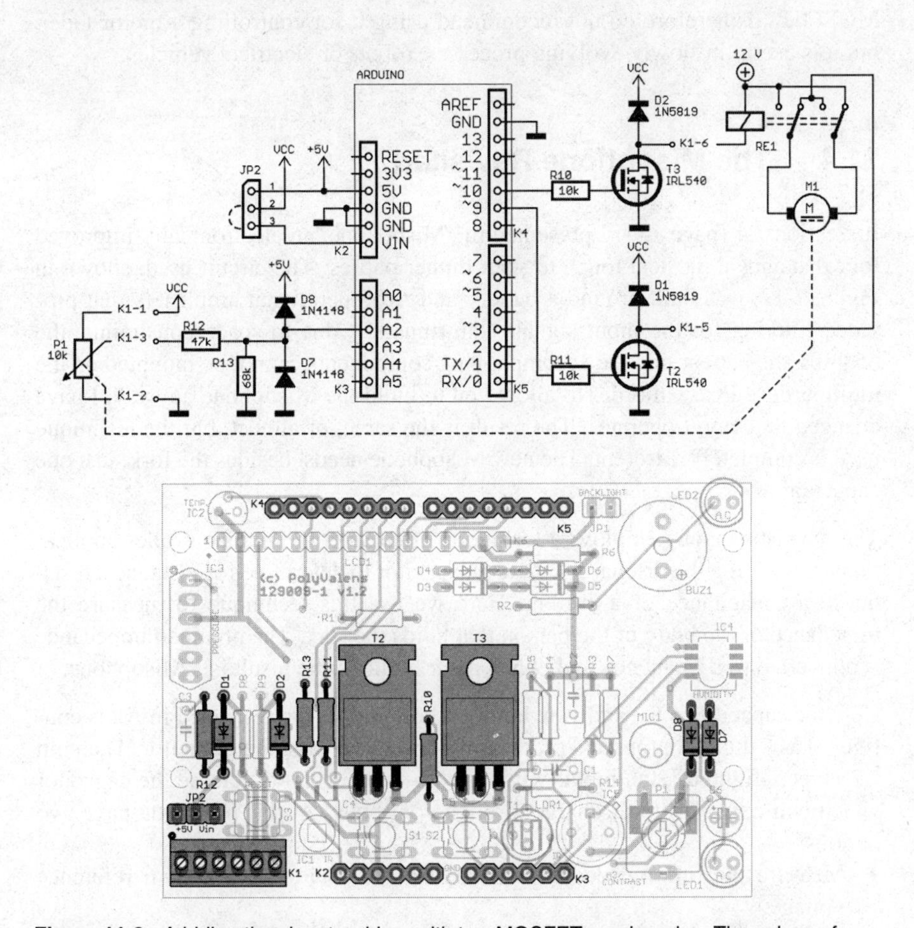

Figure 11-3 - A bidirectional motor driver with two MOSFETs and a relay. The values of R12 and R13 are calculated for a V_{cc} of 12 V. R10 and R11 allow the use of bipolar transistors instead of MOSFETs.

The analogue input of K1 can be used to connect the slider of the motor-controlled potentiometer, but it is also possible to use P1 instead (which, of course, breaks the control loop). Note that the analogue input of K1 (pin 3) is equipped with a resistive voltage divider (R12 and R13) and protection diodes (D7 and D8) and even has a filtering capacitor (C3). This allows you to measure voltages that are higher than the Arduino power supply of 5 V.

The inconvenience of this hybrid motor driver is its speed, or rather its lack thereof, because the electromechanical relay is slow, especially compared to the MOSFETs. I therefore do not recommend using it for controlling a motor fader, but it is useful in slowly evolving processes, robots or electrical vehicles.

11.4 The Misophone Revisited

In Section 7.4 (page 144) I presented the Misophone, an electronically improved fork that adds a musical touch to your dinner parties. The circuit used, shown in Figure 7-12, was built around a battery and two operational amplifiers that provided filtering for the input signal. Unfortunately, due to space constraints, the opamps, the battery and the accompanying components cannot be mounted on the multipurpose PCB. In order to allow you to build the Misophone anyway, I have changed the way it operates. The result is the same, or almost, but the technique used is completely different. The new Misophone needs, besides the fork, just one capacitor.

The new Misophone employs a technique often used in microcontroller applications: measuring the discharge rate of a capacitor to determine for instance a resistance, a capacitance or a current. Here we use this technique to measure the impedance of the body of the person that holds the fork. The obtained impedance is converted to a frequency and, when made audible, the result is a Misophone.

Here the capacitor is C3 and you, holding the fork, are the resistor. Pin A2 is configured as a digital output and set to a high level to charge the capacitor. Then pin A2 is reconfigured as an input (without pull-up resistor!) to release the capacitor which will now start to discharge through your body. At this point you have two options:

- Measure the time needed to discharge the capacitor to a known reference voltage;
- Measure the voltage left over the capacitor after discharging it during a well-defined time period.

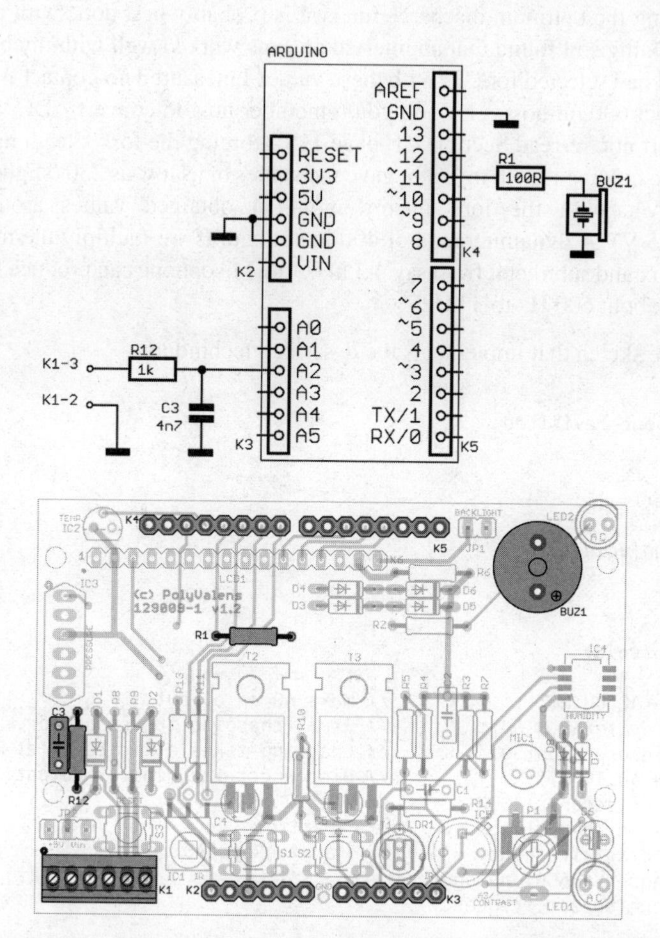

Figure 11-4 - The Misophone revisited: a capacitor, a fork and a buzzer controlled by a simple sketch. Resistor R12 protects the user against potentially dangerous currents.

The first technique can be used with the AVR's analogue comparator which is – unfortunately – not supported by Arduino; the second technique fits better in our programming environment. In both cases you will end up with a result, a time or a voltage difference, of which the value depends on the resistance of your body. You can use this value to calculate the exact resistance if you like, but for our application we can stop here and convert the value into something useable by the function `tone`.

Determining the optimum discharge interval is probably best done with an oscilloscope. Doing so I found that an interval of 2 ms worked well with my body and the value I had selected for C3. With these values I measured no-contact ADC values of about 630 (almost 3.1 V; do you remember how to convert ADC values to voltages? If not, re-read Section 7.1, page 115); forcing the fork's teeth in my left hand while holding it with my right gave me values of as low as 230 (slightly more than 1.1 V). With the fork in my mouth I obtained values around 300 (almost 1.5 V). A dynamic range of 400 is not bad. If we multiply the measured value by six and subtract it from say 4.3 kHz, the Misophone can produce frequencies from about 600 Hz to 3 kHz.

Here is the sketch that implements the described technique:

```
/*
 * misophone revisited
 */

int buzzer = 11;

void setup(void)
{
}

void loop(void)
{
  pinMode(A0,OUTPUT);             // Make A0 an output.
  digitalWrite(A0,HIGH);          // Start charging the capacitor.
  delayMicroseconds(50);          // Charging takes a little time.
  pinMode(A0,INPUT);              // Disconnect the digital output.
  delayMicroseconds(2000);        // Wait 2 ms.
  // Measure the voltage over the capacitor.
  unsigned int frequency = 4300 – 6*analogRead(A0);
  if (frequency<600) noTone(buzzer);
  else tone(buzzer,frequency);
}
```

Let's end this section with a remark on safety. The original Misophone used a battery to power the sensor (the fork); the signal conditioning circuit did not provide power to the sensor and it was therefore possible to power the rest of the circuit – including the Arduino board – from a power supply (even though it is always preferable to power circuits connected to a living body from a battery). In the revisited Misophone on the other hand the sensor is directly powered from the Arduino power supply. A current limiting resistor is available on the multipurpose PCB (R12), but still it is very much recommended to power this Misophone only from a battery (9 V for instance).

11.5 Visualize Your Data

It is always handy to have an LCD connected to a microcontroller. The circuit from Figure 8-1 (page 159) was therefore ported to the multipurpose PCB, but, of course, not one on one as that would have been too easy (for you). To save an MCU pin I replaced the backlight switching transistor by a simple jumper. Also, for PCB lay-out reasons, I wired the LCD differently. Of course this does not present any difficulties for you now that you have become the Master of MCU and therefore I will not explain here how to make the sketch from Section 8.1.1 (page 160) work. Just re-read Section 8.1.1 carefully and pay attention to the details. I am sure you will have the LCD up and running in a jiffy[1].

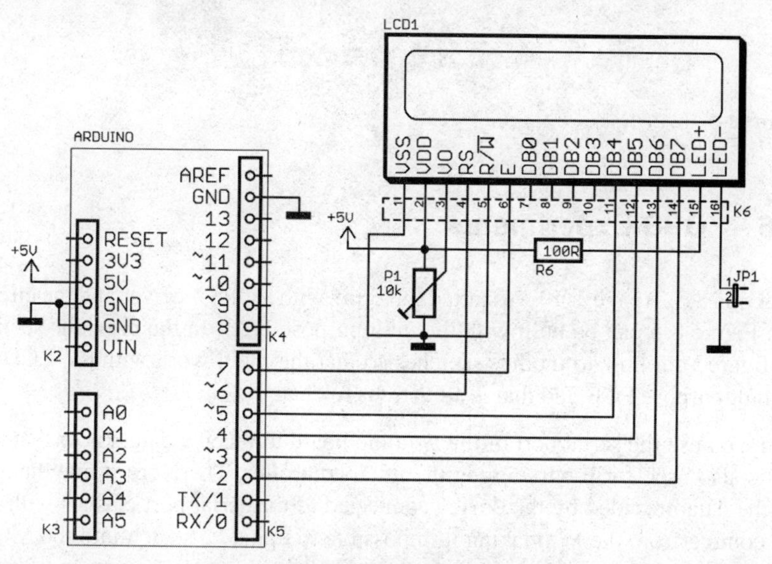

Figure 11-5 - An LCD too can be fitted to the multipurpose PCB. P1 controls the display's contrast; place a jumper on JP1 to activate the backlight. (1/2)

1. Did you find `LiquidCrystal lcd(6,7,5,4,3,2)`? Well done!

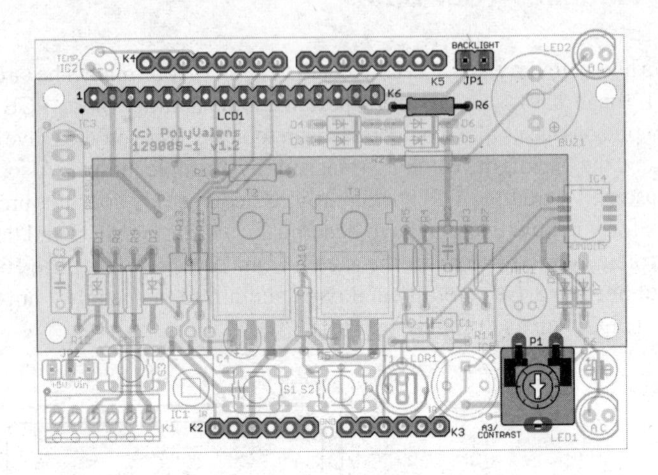

Figure 11-5 - (2/2)

11.6 GPS Experiments

In Section 8.3.2 (page 168) we started tinkering with a GPS receiver and the circuit from Figure 8-5 can be built with the multipurpose PCB. In the previous section you figured out how to modify sketches so that they will work with the LCD on the multipurpose PCB and that is all there is to change.

When you use the serial port of the Arduino board to receive data, know that this port is also used for programming the microcontroller. Therefore it may happen that the data provided by the device connected to the serial port clashes with the data coming from the PC making it impossible to upload a sketch into the MCU. If loading a sketch seems unusually long and ends with an error message like:

```
avrdude: stk500_getsync (): not in sync: resp = 0x00
```

then disconnect the device from the serial port. This should solve the problem.

R9 resistance isolates the PC data stream from the serial device's data stream, which allows you in most cases to program the micro without disconnecting the device as long as it is not sending data.

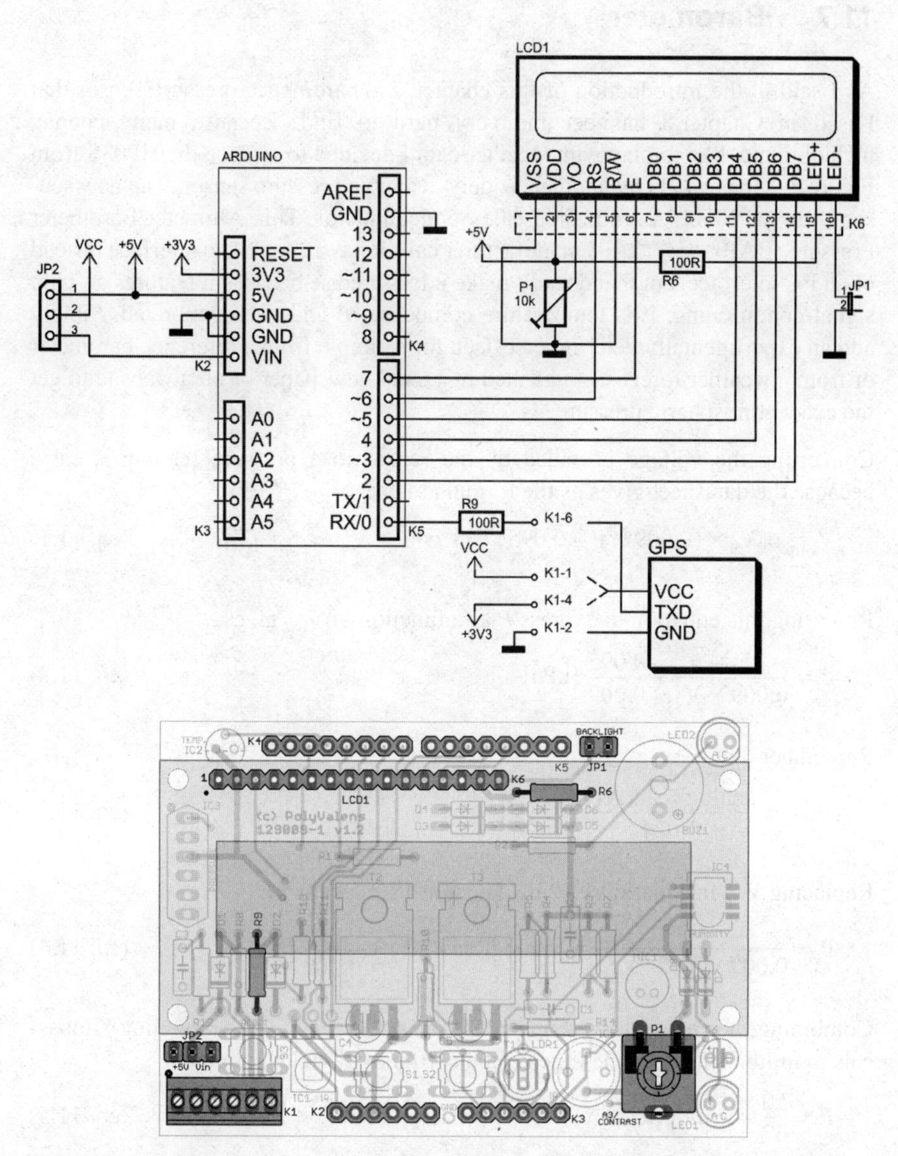

Figure 11-6 - Connector K1 allows easy connection of a GPS receiver module. Depending on the module chosen it can be powered from 5 V or 3.3 V.

11.7 Barometer

As I said in the introduction of this chapter, the barometric pressure sensor that I used in Chapter 8 has become very hard to find. Because many people, including me, like to play with weather data I decided to replace the HP03S from Figure 8-7 (page 191) by another, widely available pressure sensor. The new sensor is an MPX4115A from Freescale Semiconductor. This Altimeter/Barometer Pressure (BAP) sensor as its manufacturer calls it has an analogue interface instead of an I²C interface, but that doesn't make it less precise because it features on-chip signal conditioning, it is temperature compensated and even calibrated. Simply add an experimentally determined offset, for example from a reference barometer or from a weather forecasts published in a local newspaper or on a website to get the exact atmospheric pressure.

Converting the voltage provided by the sensor to a pressure reading is easy, because the datasheet gives us the formula to use:

$$V_{out} = V_s \times (0{,}009 \times P - 0{,}095) \text{ [V] where } V_s = 5{,}1 \text{ [V]} \qquad \text{(eq. 11.1)}$$

Rewriting this equation to express P as a function of V_{out} gives:

$$P = \frac{V_{out}}{0{,}009 \times V_s} + \frac{0{,}095}{0{,}009} \text{ [kPa]} \qquad \text{(eq. 11.2)}$$

Remember that:

$$V_{out} = \frac{\text{analogRead(A1)}}{1\,024} \times 5 \text{ [V]} \qquad \text{(eq. 11.3)}$$

Replacing V_{out} in equation 11.2 by equation 11.3 results in:

$$P = \frac{5}{0{,}009 \times 1\,024 \times V_s} \times \text{analogRead(A1)} + \frac{0{,}095}{0{,}009} \text{ [kPa]} \qquad \text{(eq. 11.4)}$$

Combining the constants by using the given value for V_s and converting kilopascals to milibars at the same time (1 kPa = 10 mbar), yields (almost):

$$P = \frac{50}{47} \times \text{analogRead(A1)} + 106 \text{ [kPa]} \qquad \text{(eq. 11.5)}$$

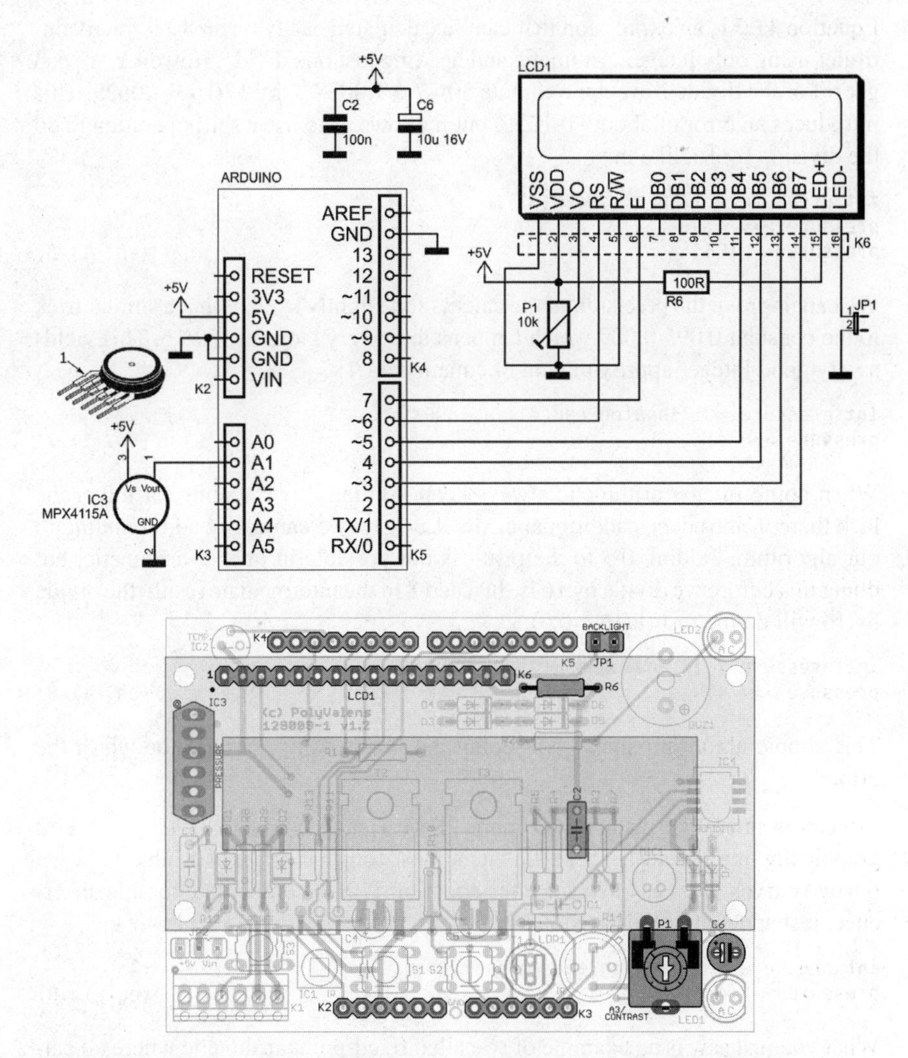

Figure 11-7 - Thanks to its analogue interface connecting the MPX4115A pressure sensor to a microcontroller is very easy. Note that the sensor exists in several variants with different casings (and suffixes). They should all fit on the PCB, but only the one in the 867-08 case will fold under the LCD.

Equation 11.5 is an expression that can be calculated easily on an 8-bit microcontroller using only integer arithmetic and needing just one divide. However, we can get rid of the divide if we approximate 50/47 = 1.0638... by 17/16 = 1.0625. This introduces an error of about –0.125% but it allows us to use a shift operation to do the division by 16, like this:

```
int pressure = 17*analogRead(A1);
pressure >>= 4;
pressure += 106;                                              (eq. 11.6)
```

We can improve the precision of the calculation slightly by applying a similar trick to the constant 0.095/0.009 which happens to be very close to 169/16. This yields a very good integer approximation of equation 11.1:

```
int pressure = 17*analogRead(A1) + 1690;
pressure >>= 4;                                               (eq. 11.7)
```

When doing integer arithmetic always keep in mind that rounding, or rather the lack thereof, introduces additional errors. Luckily, we can easily add rounding to our algorithm. Adding 0.5 to the result is not possible in integer arithmetic, but doing this before we divide by 16 is. Just add 8 to the intermediate result, the divide by 16 will do the rest (8/16 = 0.5):

```
int pressure = 17*analogRead(A1) + 1698;
pressure >>= 4;                                               (eq. 11.8)
```

This simple algorithm shows that doing the maths can sometimes be worth the effort.

We can improve the results a bit more. If we visualize equations 11.2 and 11.8 graphically in a spreadsheet, we find an almost constant offset of 1 mbar between the two curves. It is easy to change the equation 11.8 to compensate for this difference, just add 1 after the division by 16 or 16 before the division, like this:

```
int pressure = 17*analogRead(A1) + 1714;
pressure >>= 4;                                               (eq. 11.9)
```

What you just saw is an example of so-called fixed-point arithmetic where all calculations are done with scaled values and where the result must be scaled down to obtain the real result. Scaling can be done with powers of ten (10, 100, 1000, etc.), but any suitable value can be used. For binary systems powers of two are useful because of the shift operators (left and right).

Fixed-point arithmetic is a good compromise for doing relatively precise calculations on platforms that are not too well equipped for efficient floating-point arithmetic. When doing fixed-point arithmetic make sure that the scaled-up values fit

in the data type used for the variables. In the previous example a 16-bit data type is enough because the maximum value possible before scaling down is $17 \times 1{,}023 + 1{,}698 = 19{,}089$ or 0x4a91 in hexadecimal. As you can see, this value is only 15 bits wide.

11.8 Humidity and Temperature Meter

The hygrometer from Figure 8-11 (page 204) can be built on the multipurpose PCB without any changes; all you have to mount are IC4, C2, C6 and R7. The sketch from Section 8.5.2 (page 202) only needs the LCD modification from Section 11.5 to display the data.

Figure 8-11 also includes a pressure sensor. I have shown you in Section 11.7 how to replace it by a different sensor. It is a good exercise to try to combine the previous section with this one to create a small weather station like the one presented in Section 8.6 (page 209).

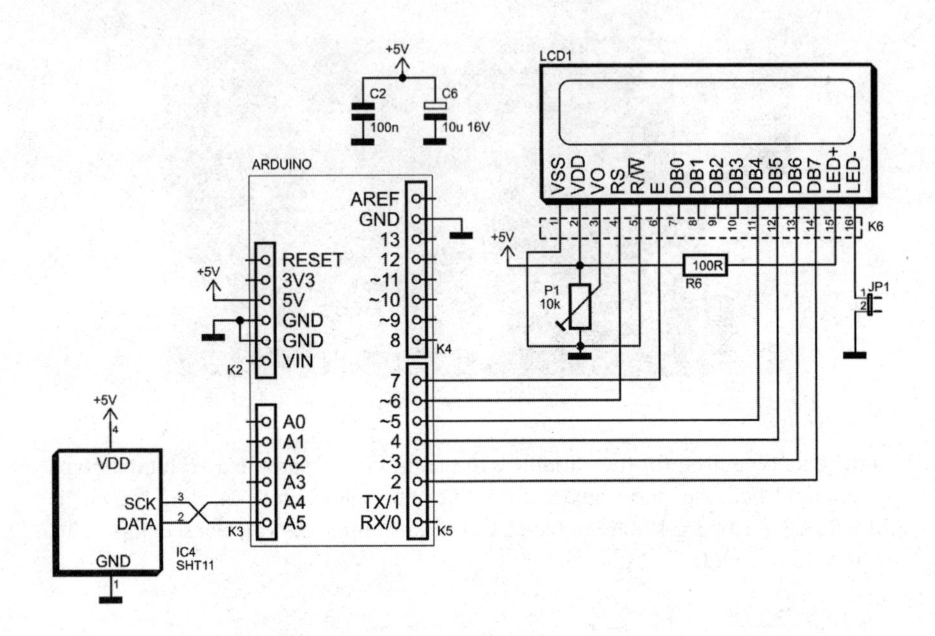

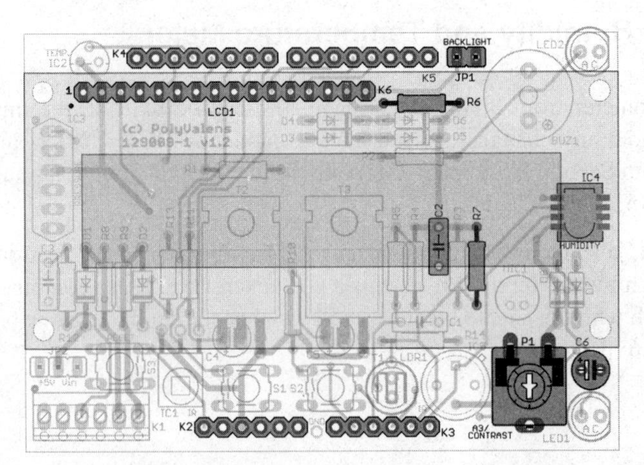

Figure 11-8 - The sensor SHT11 (IC4) measures temperature and humidity.

11.9 DCF77 Receiver

In Section 9.1.1 (page 222) we connected a DCF77 receiver to the Arduino board to help us learn how to handle time and time related functions on a microcontroller. The circuits from Figures 9-2 and 9-3 can be built with the multipurpose PCB. In Section 11.5 you discovered how to modify the sketches so that they will work with the LCD on the PCB and that is all there is to change.

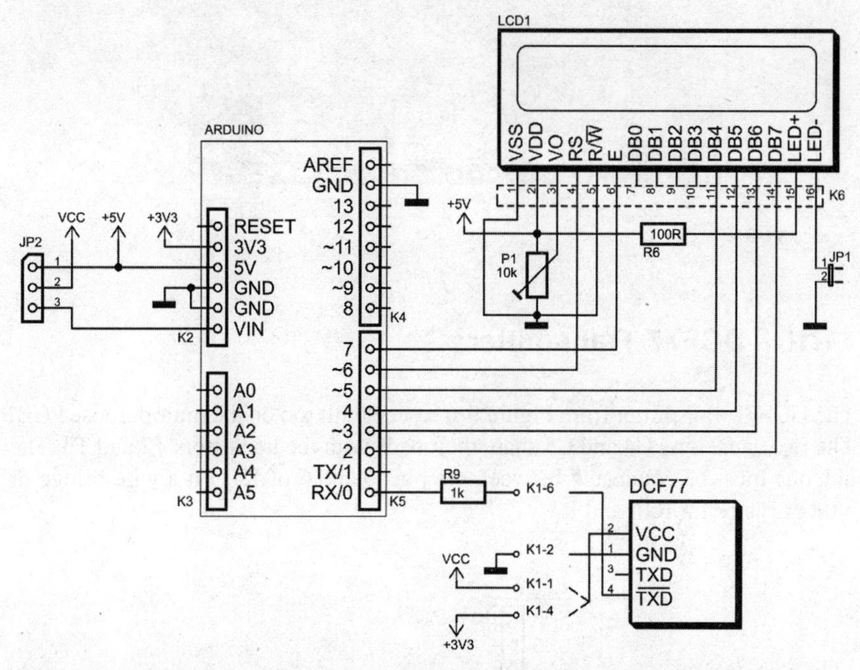

Figure 11-9 - Connector K1 allows easy connection of a DCF77 module. The module can be powered from 5 V or 3.3 V and three Arduino inputs are available: A2, Pin 0 and Pin 1. Note that the circuit presented here requires a further modification of the sketch to adapt it to the new input. (1/2)

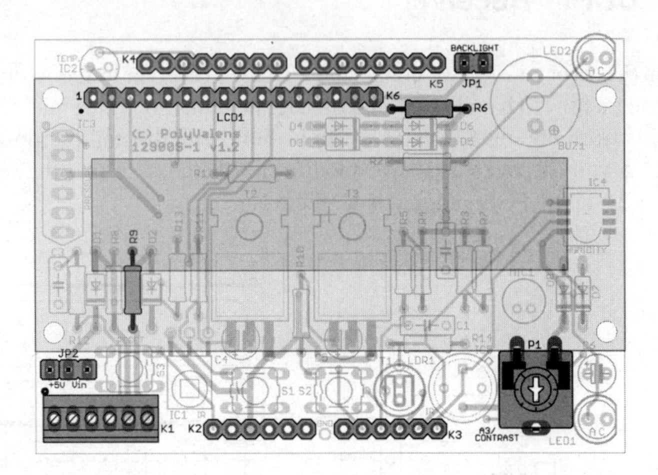

Figure 11-9 - (2/2)

11.10 DCF77 Transmitter

The DCF77 transmitter from Figure 9-6 can be built too on the multipurpose PCB. The two capacitors C4 and C5 share their pads with the transistors T2 and T3. The antenna must be connected between the pins 5 and 6 of K1. Fit a wire bridge or a 0 Ω resistor for R10 and R11.

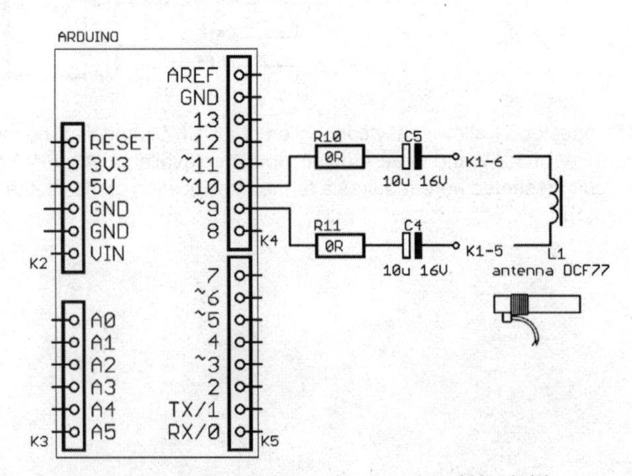

Figure 11-10 - Connect the antenna between pins 5 and 6 of K1. (1/2)

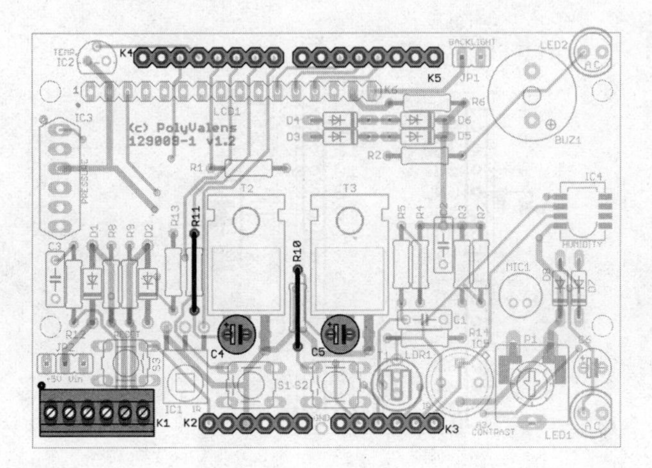

Figure 11-10 - (2/2)

The DCF77 power-line transmitter from Figure 9-7 is possible if you fit wire bridges in place of R11 and C4. Pin 5 of K1 is the output to control the intercom.

11.11 Infrared Receiver

In Section 9.8.1 (page 250), Figure 9-11, I showed you that it is possible to use MCU pins to power a sensor, an infrared receiver in this case. In order to save MCU pins on the multipurpose PCB I decided to wire the power pins of the sensor directly to the power lines, only the sensor's signal pin is routed to the MCU. This means that it is possible to simplify the functions `setup` a bit in the corresponding sketches, because you don't have to worry anymore about powering the sensor and filtering out spurious pulses. The part should be up and running when the sketch starts.

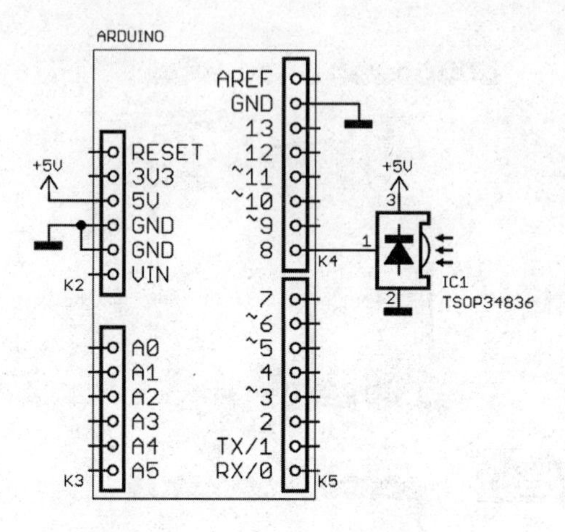

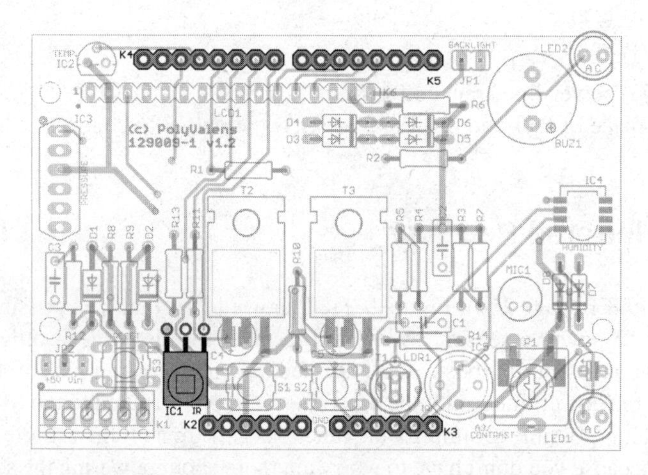

Figure 11-11 - The IR receiver is no longer powered by the microcontroller.

11.12 Infrared Transmitter

Because on the multipurpose PCB the LCD occupies Arduino's pin 2 and because I thought it would be nice to be able to use the LCD together with the infrared LED, to make a sophisticated remote control for example, the infrared LED from Figure 9-14 (page 270) is connected to another pin, number 13. However, the PCB can also have an infrared LED on pin 11, so actually you have two options to make an infrared transmitter or make it a two-channel type.

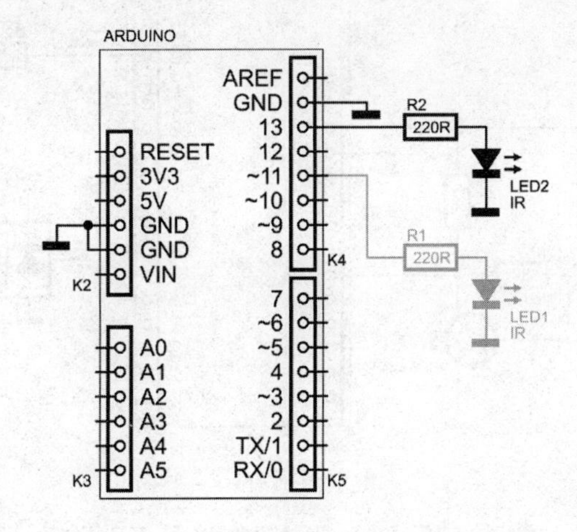

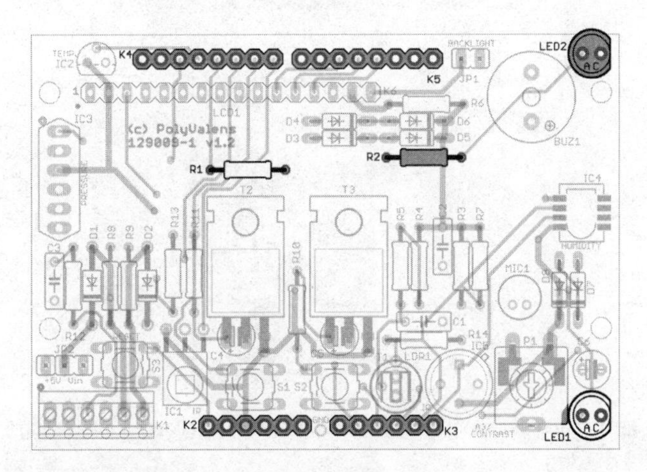

Figure 11-12 - The IR transmitter can have two channels.

11.13 Rioter

The circuit from Figure 9-15 (page 277) on the multipurpose PCB has the same infrared LED options as the infrared transmitter from the previous section. The rest of the circuit is unchanged, so the corresponding sketch only needs a very small change to make the LED work. I leave this exercise to you, master of MCU.

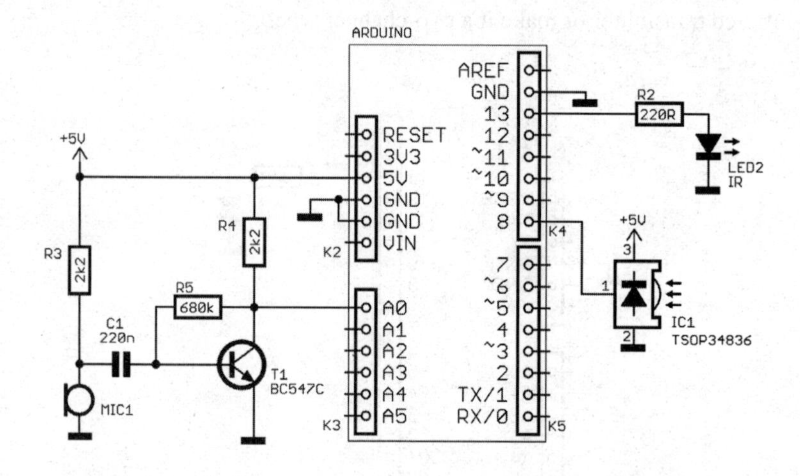

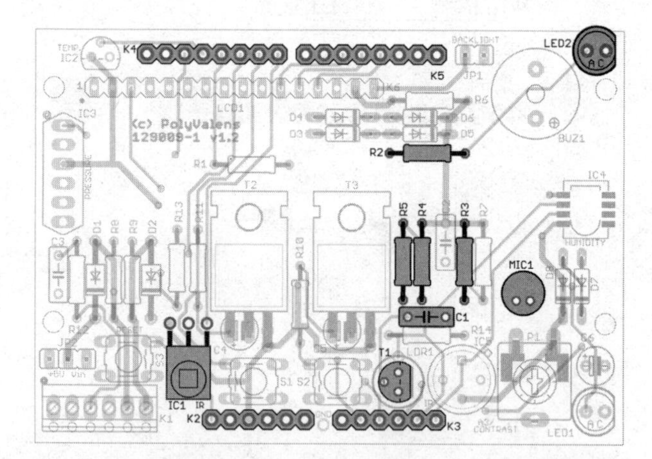

Figure 11-13 - A sound detector and an infrared transceiver allow for interesting applications.

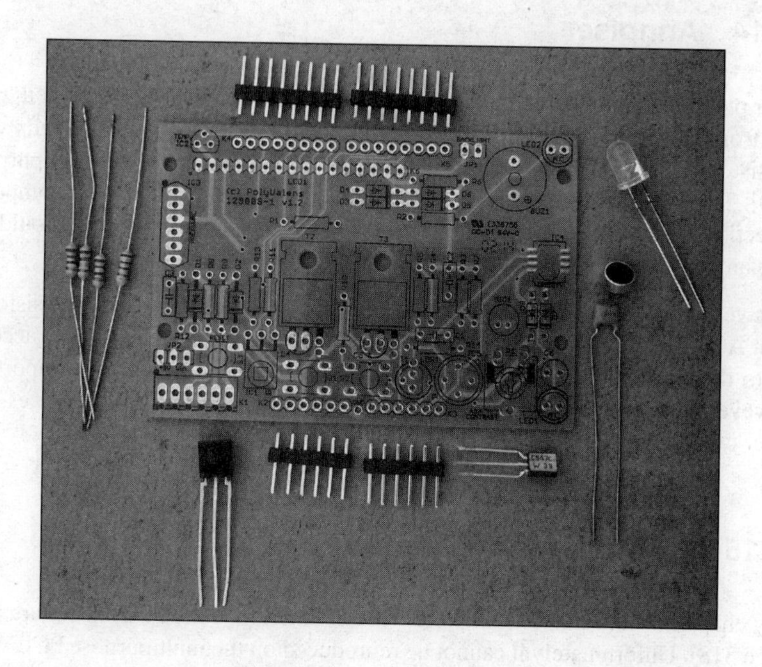

If the range of the transmitter is insufficient, you can increase the current through the infrared LED. However, be very careful to never exceed 40 mA, the maximum allowed current for an output of the MCU. If you go beyond this maximum, you risk damaging the MCU. Moreover, the datasheet of the AVR stops at a maximum current of 20 mA for all specifications and if I was you I would do the same. Also do not forget that the LED2 is in parallel with the LED on the Arduino board that also consumes a bit of power. If this is inconvenient, then use the LED1 of the multipurpose PCB as infrared emitter. Another possibility is to use one of the two power transistors that can be mounted on the PCB. In this case it is the infrared LED that determines the maximum allowed current (consult its datasheet). Don't forget to invert the LED's signal to cancel the inversion introduced by the transistor.

11.14 Annoiser

After playing with invisible light we turn our attention to visible light with the circuit from Figure 10-5 (page 314). The photoresistor LDR1 shares its pads with transistor T1, meaning that you cannot mount both together on the multipurpose PCB[1]. The photoresistor is now connected to a different analogue input compared to Section 10.10.1 (page 313), so you must adapt the sketch. Again, you should not have any problems of doing this all by yourself.

To save a resistor, you can replace resistor R4 by the integrated pull-up resistor of input A0, even though its value is not well defined (between 20 and 50 kΩ according to the MCU's datasheet). In this application its value is not very important. However, if you must ensure the exact same operation for say 10,000 devices, it is better to use an external resistor with a known value.

11.15 La Cucaracha in Stereo

This chapter would not be complete without the noisy circuit from Figure 10-7 (page 318). Unfortunately it cannot be reproduced on the multipurpose PCB without some changes. Also, the relay does not fit on the PCB. The temperature sensor is connected to a different pin, like the photoresistor, and you have two options for the transistor that switches the relay. This means that you can control two car horns instead of just one! The LED indicator that shows the state of the relay has gone, but you can use one of the on-board LEDs – or both – for this function.

Of course you will have to modify the sketch from Section 10.11.1 (page 319) to reflect the changes made to the circuit. The connection of the temperature sensor has been simplified, like those of the infrared receiver from Section 11.11, meaning that the function `setup` can be simplified too. The sensor's output is now connected to Arduino pin 12 instead of pin 4. This means that the definitions (`#define`) for the DQ pin of the sensor at the stop of the sketch must be changed to use port B instead of port D. This is quite simple, just replace 'D' with 'B' like this: DDRD becomes DDRB, PORTB becomes PORTD and PIND becomes PINB. The other code modifications needed are also very simple and you can figure them out yourself. If you want to control two relays instead of one, you will have to add a few lines of code. Let the sketch inspire you, all you need to know to do this is in there.

1. You could replace R5 by LDR1 to make a light-dependent amplifier. If you do this, you must mount T1 too.

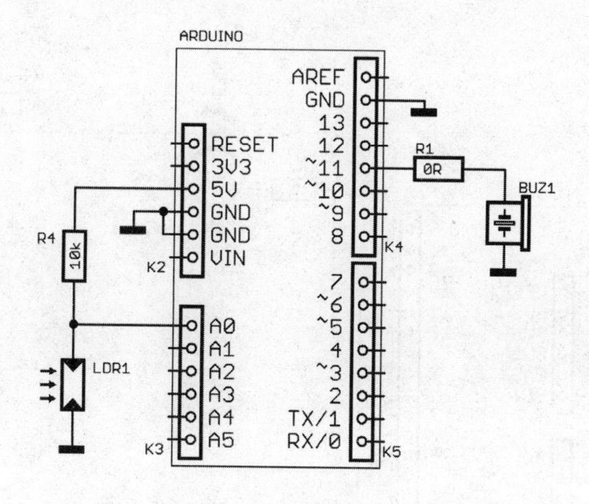

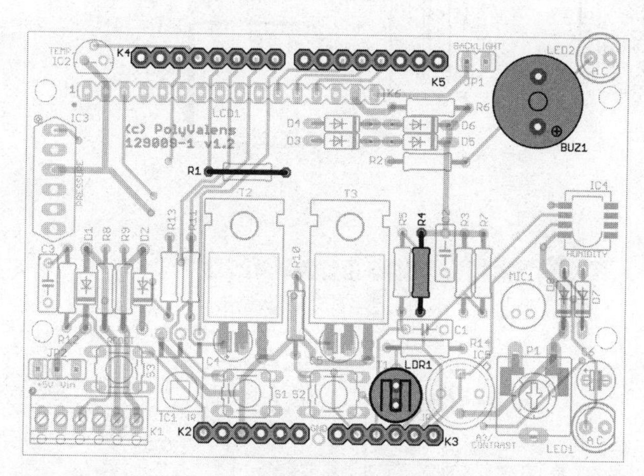

Figure 11-14 - Make sure your sketch uses the correct analogue input for photoresistor LDR1.

If you can't manage to load the sketch into the MCU, disconnect the relay (K1-6). Chances are that it disturbs the serial port's input, rendering programming of the MCU impossible.

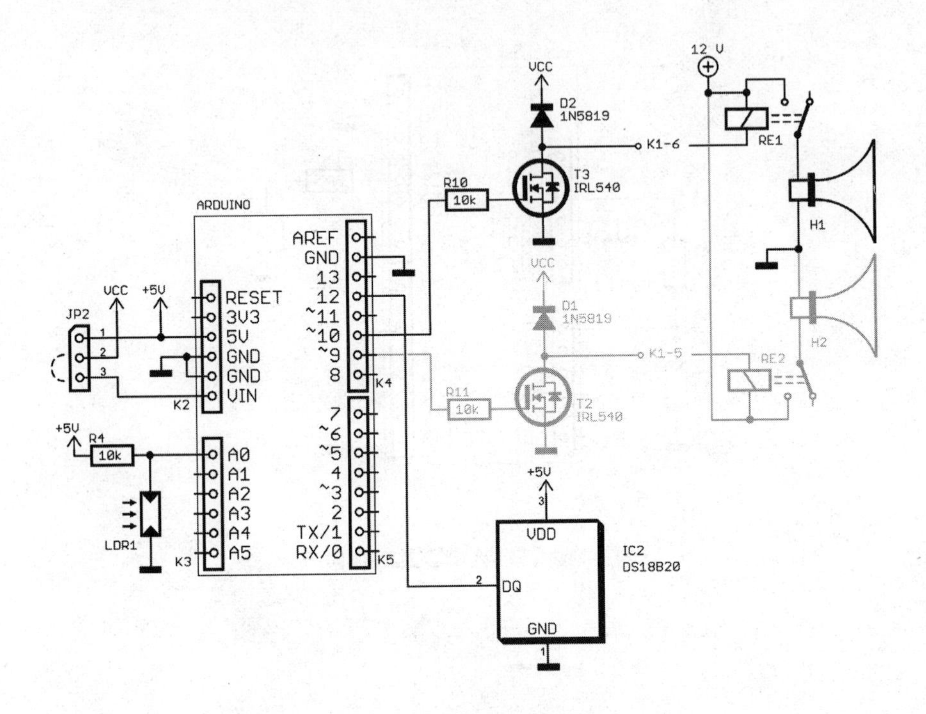

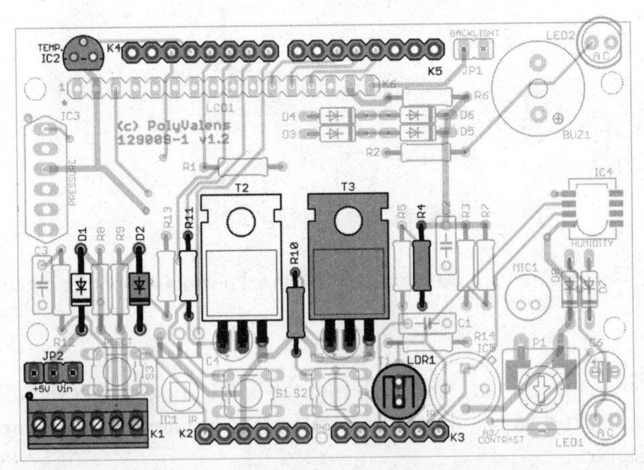

Figure 11-15 - With two transistors you can control two relays that in turn each control a horn. The result: a very loud alarm in full stereo! The components drawn in grey in the schematic correspond to the white parts on the PCB. Note that the 12 V used to power the horn(s) and relay(s) can also be used to power the Arduino board. In this case place a jumper on pins 2 and 3 of JP2.

11.16 Fire Detector

The last circuit from the previous chapters that you can build with the multipurpose PCB is the fire detector from Figure 10-9 (page 326). The sensor IC5 with its SMBus interface is connected in the same way as in Section 10.12.1 (page 325); refer to the Section 11.15 for the output stage.

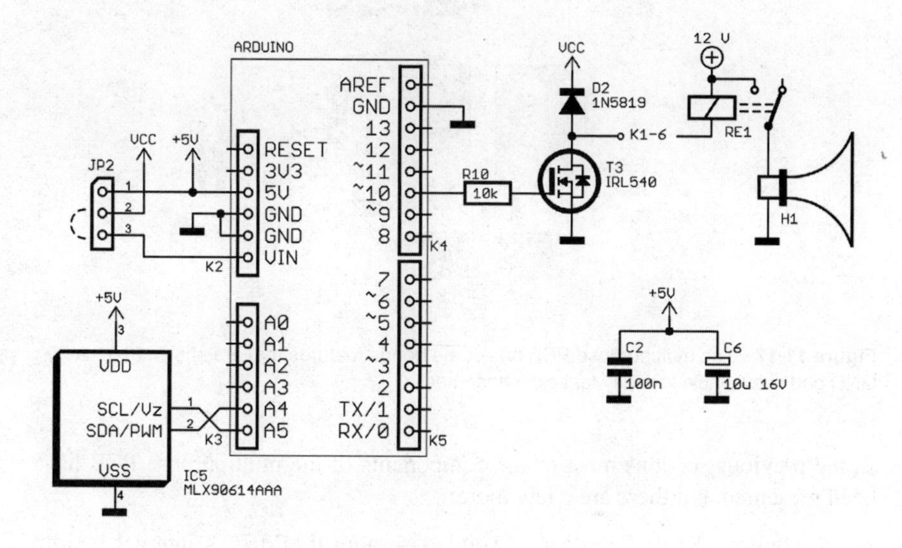

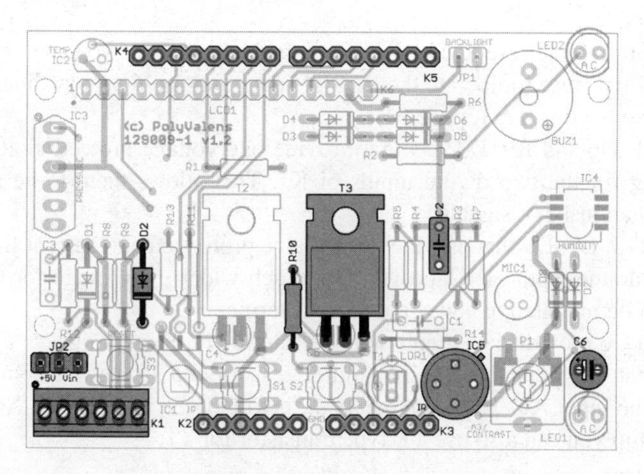

Figure 11-16 - Save homes and lives with this fire alarm. Beware, IC5 is also available as a 3.3 V device, known as MLX90614<u>B</u>AA.

11.17 Bonus

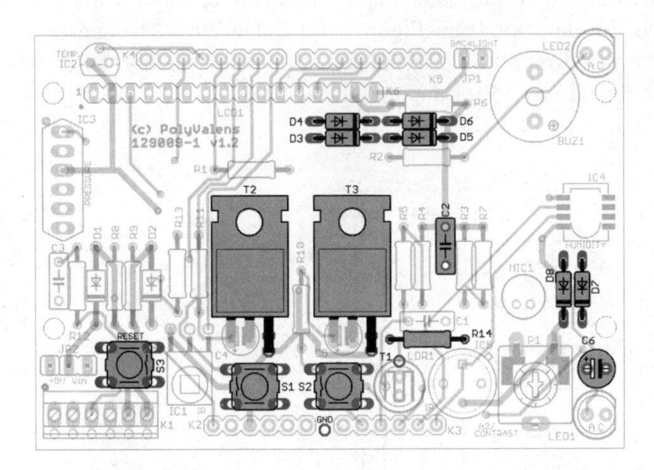

Figure 11-17 - The multipurpose PCB has some bonus features that are shown here. A GND pad is available for easy test clip attachment.

In the previous sections most of the components of the multipurpose PCB have been presented, but there are a few more:

- Pushbuttons S1 & S2, always useful. Use with the MCU's internal pull-up resistors;
- Pushbutton S3 to reset the MCU. If you feel adventurous you can program the RSTDISBL fuse to liberate the RESET input and use it as a pushbutton input (see also Section 4.5.1, page 54);
- R8, D4, D6 and R9, D3, D5 that provide overvoltage protection and current limiting for the two digital inputs of K1. This allows them to be used with more than just 5 V signals;
- R12, R13, D7, D8 that provide overvoltage protection and current limiting for the analogue input of K1 providing a much wider input range for this input. C3 can be mounted to add some noise filtering;
- C2 and C6 provide 5 V supply voltage filtering or decoupling;
- R14 that can play the role of emitter resistor for T1 with a connection to analogue input A4. The extra pad *pad2* is to be used in this case. Note that in this setup you can also use a P-type transistor for T1;

- T2 and T3 both have an extra emitter/source pad so you can mount P-type transistors instead of N-type. Using these pads make them switch to +5 V instead of to 0 V;
- R10 and R11 allow the use of bipolar transistors for T2 and T3 instead of MOSFETs.

Refer to Figure 11.1 for the full schematic of the multipurpose PCB. Study it closely and let your imagination soar; combine circuits and sketches; replace parts. Imagination is the beginning of creation. You imagine what you desire; you will what you imagine; and at last you create what you will[1]. After all, you are the Master of MCU.

1. George Bernard Shaw

Programs Overview

Chapter 5

Chapter 6

Chapter 7

Chapter 8

Chapter 9

Chapter 10

Chapter 11

Illustrations Overview

Tables Overview

Symbols

Numerics

A

B

C

J

K

L

M

Q

R

S

T

U

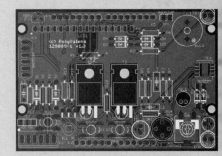

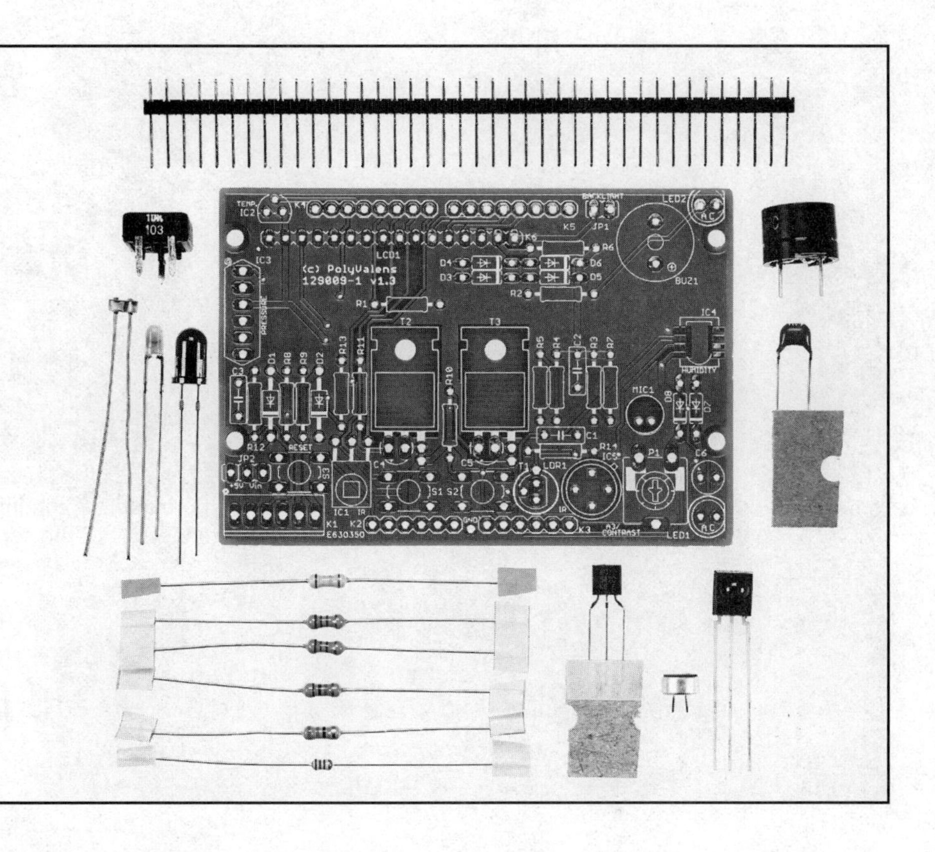

LDR1/R14 = LDR 200 kΩ

P1 = potentiometer 10 kΩ

• Capacitor

C1 =220 nF, X7R, 50 V

• Semiconductors

LED1 = LED, green, 3 mm

LED2 = IR LED, 940 nm, 5 mm

T1 = BC547C

IC1 = TSOP34836 (IR receiver)

• Various

MIC1 = electret microphone, 6 mm

BUZ1 = buzzer 12 mm

40-pin header

Order Today at www.elektor.com/arduino-multipurpose-shield-kit-129009-71